Preface

This book is designed to enable you to review in just 1-2 days all of the basic microbiological sciences you studied in the first year of medical school: Bacteriology, Virology, Immunology, Parasitology and Mycology. We have been able to condense a review of all basic microbiological sciences into a single book because of the new format of the National Board Part I Exam. It is no longer prudent to review exhaustively the basic science courses because the new examination format no longer rewards an encyclopedic knowledge of the basic sciences. Instead, the new exams test knowledge of the scientific basis of disease and the ability to apply basic scientific information to the clinical reasoning process. Consequently, the most efficient way to study for the new exam is 1) to review only the most clinically relevant material from each basic science course and 2) to focus on the application of this material to the solution of clinical problems. These two new study features form the core of this text.

If you answer every question and read all the tutorials in this book, you can cover within 2 days all of the most clinically relevant information from your basic microbiological science courses. You will find that many microbiological facts reviewed or learned anew will be presented in the context of a clinical case or an illustration. We hope that the clinical cases and illustrations will enhance your understanding and recall of the information. Finally, you will learn from the tutorials how microbiological information is used by knowledgeable physicians to understand the courses of diseases and the significance of abnormal findings.

Gerald V. Stokes, Ph. D.
Washington, D.C.
March, 1993

Acknowledgements

The author would like to thank Kurt E. Johnson Ph. D., Professor, Department of Anatomy, The George Washington University Medical Center for his encouragement on tackling this project and his skilled editorial comments and direction in formatting the final document for publication. He would also like to thank Melvin Reich, Ph. D., Professor, Department of Microbiology and Immunology, The George Washington University Medical Center for his critical review of the final manuscript. Finally, the author would like to thank Ronald C. Bohn, Ph. D., Associate Professor, Department of Anatomy, The George Washington University Medical Center for his skillful manipulation of the final documents for publication. The author is grateful for all of this help and acknowledges any errors in this book as his own.

Disclaimer

The clinical information presented in this book is accurate for the purposes of review for licensure examinations but in no way should be used to treat patients or substituted for modern clinical training. Proper diagnosis and treatment of patients requires comprehensive evaluation of all symptoms, careful monitoring for adverse responses to treatment and assessment of the long-term consequences of therapeutic intervention.

Figure Credits

Figures 1.1, 3.2, 3.3, 4.2, 4.3 and 4.4. From J.C. Sherris, ed., Medical Microbiology, An Introduction to Infectious Diseases, ed. 2, ©1990 reprinted by permission of Appleton & Lange, Norwalk, CT. and John C. Sherris. Figures 4.1, 4.3 and 4.4 were drawn by Sam C. Eng.

Figure 4.1. From Morbidity and Mortality Weekly Report, February 1, 1991. Reprinted by permission of Centers for Disease Prevention and Control.

Table of Contents

Dedication

Gerald V. Stokes would like to dedicate this book to Charlotte, Gordon and Garrett who remained patient while he assembled this collection of concepts.

J&S

Composition and Layout: Ronald C. Bohn, Ph. D.
Cover Design: Kurt E. Johnson, Ph. D.
Printing Supervisor: Robert Perotti, Jr.
Printing: Goodway Graphics, Springfield, Virginia

Library of Congress Catalog Card Number 93-077103

ISBN 0-9632873-2-X

P 10 9 8 7 6 5 4 3 2

Microbiology:
Review for New National Boards

Gerald V. Stokes, Ph. D.
Associate Professor of Microbiology and Immunology
George Washington University Medical Center
Washington, D.C.

J&S

J&S Publishing Company Inc., Alexandria, Virginia

CHAPTER I

BACTERIOLOGY

Items 1-4

A week-old newborn was admitted to the intensive care unit after an apparently normal delivery from a healthy mother. The infant had displayed abnormal eating patterns and reduced levels of behavioral alertness indicative of neurological involvement. The routine bacteriological cultures of both blood and CSF were negative. CSF plated on Eaton's agar yielded minute slow-growing colonies.

1. The observed bacteremia is most likely due to an infection caused by bacteria consisting of

    (A) gram-positive organisms
    (B) gram-negative organisms
    (C) gram-variable organisms
    (D) non-gram-stainable organisms
    (E) a mixed infection

2. Speciation of the causative agent(s) could be most accurately determined by

    (A) microscopic examination
    (B) carbohydrate fermentations
    (C) immunofluorescence using monoclonal antibody to surface antigens
    (D) growth on differential media
    (E) colony characteristics

3. An antibiotic generally prescribed in the treatment of this type of neonatal infection is

    (A) penicillin
    (B) tetracycline
    (C) chloramphenicol
    (D) erythromycin
    (E) gentamicin

4. Which of the following organisms best fits the isolate recovered from either the blood or CSF specimen?

(A) Listeria monocytogenes
(B) Chlamydia trachomatis
(C) Neisseria meningitidis
(D) Haemophilus influenzae
(E) Ureaplasma urealyticum

ANSWERS AND TUTORIAL ON ITEMS 1-4

The answers are: **1-D; 2-C; 3-D; 4-E**. An increasing number of women are asymptomatic carriers of mycoplasmas in the genital tract. Ureaplasma urealyticum is the slowest growing of the mycoplasmas. They are isolatable on special nutrient media, such as Eaton's agar, which contains highly purified ingredients and antibiotics which prevent the overgrowth of most conventional bacteria. As with all mycoplasmas, the Ureaplasma do not contain a cell wall, thus do not react with Gram stain. The absence of a cell wall also precludes the choice of the penicillins. The organisms are surrounded only by a cell membrane, which contains species-specific antigens and others which cross-react with human antigens. Many are pleomorphic. The Ureaplasma differ from Mycoplasma by their urea requirement. All mycoplasmal organisms require sterols for growth.

There are approximately fourteen recognized serovars of Ureaplasma urealyticum. Most have invasive abilities and are not normally found in either the blood or cerebrospinal fluid of healthy people; however, there are reported cases involving the latter, as presented in this case history. Generally they are confined to the surface colonization of epithelial cells of the genital tract of infected individuals. The genus Ureaplasma is easily distinguished by the presence of unique membrane antigens detectable using immunofluorescence and monoclonal antisera.

Erythromycin is an effective antibiotic in the treatment of Ureaplasma infections. It is active in the inhibition of bacterial protein synthesis, as is tetracycline, but without the deleterious side effects on developing bones in the newborn.

Items 5-8

A male patient with culture-positive gonococcal urethritis was treated with 5 million units of aqueous procaine penicillin G intramuscularly plus probenecid (1.0 g) orally. Eight days later, he returned complaining of similar symptoms consisting of painful urination, but less urethral discharge. A second pathogen was suspected.

5. The symptoms expressed in the patient indicate that

   (A) reinfection should be suspected
   (B) he may have nongonococcal urethritis
   (C) the bacterial strain has acquired a drug resistance plasmid
   (D) he has a viral infection
   (E) he has a fungal infection

6. IF the causative agent responsible for the above infection were a gram-negative diplococcus, and it had developed antibiotic resistance this could be confirmed by which of the following tests?

   (A) coagulase
   (B) catalase
   (C) Kirby Bauer
   (D) methyl red
   (E) methylene blue

7. The second organism most likely responsible for the above infection is

   (A) Mycoplasma pneumoniae
   (B) Chlamydia trachomatis
   (C) Candida albicans
   (D) human papilloma virus
   (E) Escherichia coli

8. An antibiotic which has been shown effective in the treatment of bacterial infections of this type is

   (A) doxycycline
   (B) ampicillin
   (C) cephalosporin C
   (D) streptomycin
   (E) amphotericin B

Items 9-12

A young college student spent his summer working for a forest service agency in Oregon. A month after returning home to Virginia, he experienced a sudden onset of fever, nausea, muscle aches and headaches. Blood and stool cultures were negative. His symptoms subsided and he assumed he had recovered. Two weeks later the symptoms reappeared, but this time accompanied with a marked splenomegaly. He also noticed a single tennis ball-sized bright red rash on his lower left leg which increased in size and sensitivity. This time a blood smear showed the presence of spirochetes.

9. The symptoms described in this case history are consistent with a preliminary diagnosis of

    (A) Rocky Mountain spotted fever
    (B) Lyme disease
    (C) syphilis
    (D) plague
    (E) leptospirosis

10. The most probable etiologic agent for the above illness would be due to organisms of the genus

    (A) Spirocheta
    (B) Rickettsia
    (C) Treponema
    (D) Borrelia
    (E) Leptospira

11. The etiologic agent responsible for the infection was most likely acquired by

    (A) a tick bite
    (B) drinking contaminated water
    (C) sexual contact
    (D) consumption of contaminated foods
    (E) inhalation

12. All of the following diseases are caused by a spirochete **EXCEPT**:

    (A) syphilis
    (B) relapsing fever
    (C) Lyme disease
    (D) leptospirosis
    (E) leprosy

ANSWERS AND TUTORIAL ON ITEMS 5-8

The answers are: **5-B; 6-C; 7-B, 8-A**. Most Neisseria gonorrhoeae isolates respond to penicillin. Multiple drug resistant strains are becoming more frequent. Therapy against such organisms would not necessarily cause a diminution of symptoms. Drug resistance can be determined diagnostically using the classical agar plate antibiotic sensitivity tests (e.g. Kirby Bauer) and minimal inhibitory concentration (MIC) determination.

The overall response is best explained by a dual infection with another organism of different antibiotic sensitivity. Sexually transmitted disease clinics report high (>40%) instances of such dual infections consisting of N. gonorrhoeae and Chlamydia trachomatis, a major agent of nongonococcal urethritis. Chlamydiae do not respond to penicillin therapy. The tetracyclines, erythromycin and azithromycin are effective in treating chlamydial infections.

ANSWERS AND TUTORIAL ON ITEMS 9-12

The answers are: **9-B; 10-D; 11-A; 12-E**. Lyme disease is a bacterial infection caused by the spirochete Borrelia burgdorferi. It is a zoonotic infection within the deer and rat populations and accidentally transmitted to man through the bite of the deer tick. The extremely small size (< 2mm) of the deer tick renders them difficult to detect. A large red petechial rash usually develops at the site of the tick bite. It may grow to over 10 cm in diameter. Different arthropod vectors are instrumental in the transmission of other Borrelia sp. infections.

Relapsing fever is another disease caused by bacteria of the genus Borrelia. Syphilis and leptospirosis are diseases also caused by spirochetes. Leprosy is caused by the gram-positive, acid-fast bacillus Mycobacterium leprae.

Items 13-18

A retired auto worker was diagnosed with tuberculosis. He had been ill for about nine months with symptoms which included intermittent fever, night sweats and a productive cough. He noted a marked weight loss of 50 lbs. A sputum smear showed the presence of numerous gram-positive acid-fast bacilli. Several of his co-workers were shown to be skin test positive with the tuberculin (PPD) test.

13. Transmission of the causative agent Mycobacterium tuberculosis between humans is primarily through

    (A) sexual contact
    (B) respiratory droplets
    (C) contaminated foods
    (D) soil (environment)
    (E) contaminated blood products

14. It would be advisable for the above patients to undergo further testing with

    (A) another skin test
    (B) a CT scan
    (C) a chest X-ray
    (D) a NMR scan
    (E) a bacterial culture isolation

15. Many virulent strains of the bacterium have been shown to contain

    (A) purified protein derivative (PPD)
    (B) small RNA plasmids
    (C) flagella and pili
    (D) cord factor
    (E) potent exotoxins

16. An example of atypical mycobacteria found frequently in conjunction with immunosuppressed patients such as AIDS patients would be

    (A) M. kansasii
    (B) M. marinum
    (C) M. smegmatis
    (D) M. scrofulaceum
    (E) M. avium-intracellulare

17. The early stage of M. tuberculosis infection is characterized by the formation of which of the following lesions in the lungs?

(A) granulomatous
(B) exudative
(C) tubercle
(D) eschar
(E) induration

18. Antibiotic therapy used in the treatment of M. tuberculosis has become more complicated by the development of multiple drug resistance. All of the following drugs, alone or in combination, have been shown effective in the treatment of the bacteria **EXCEPT:**

(A) chloramphenicol
(B) isoniazid
(C) rifampicin
(D) ethambutol
(E) pyrazinamide

ANSWERS AND TUTORIAL ON ITEMS 13-18

The answers are: **13-B; 14-C; 15-D; 16-E; 17-B; 18-A**. There has been a reemergence of tuberculosis in the United States affecting a broad spectrum of the population. This is in part due to the increase of immune deficiency diseases, multiple drug resistance and noncompliance with prescribed antibiotic therapy regimen. Mycobacterium tuberculosis, the causative agent of the disease, is acquired by the inhalation of respiratory droplets from infected individuals. Initial infection therefore takes place in the lungs with the formation of exudative lesions which progress into granulomatous and finally tubercle lesions consisting of walled-in bacteria, giant and epithelioid cells surrounded by fibrosis and calcification.

The highly immunogenic nature of the mycobacterial cell wall elicits a strong cell-mediated immune response. Purified protein derivative (PPD), administered intradermally, produces induration (thickening) and erythema (reddening) in the area of injection in previously exposed individuals. Induration in excess of 10 mm within 48-72 hours indicates a positive skin test. Virulence in the bacteria is closely correlated with the presence of cord factor (trehalose dimycolate) in the organisms. Most Mycobacteria sp. display slow growth rates. The cell-mediated immune response attempts to restrict the spread in healthy individuals, therefore immunocompromised individuals are more prone to infections. M. avium-intracellulare is an atypical mycobacterium and has been observed with high frequency in such patients.

Antibiotic therapy usually consists of a combination of isoniazid, rifampicin, ethambutol (or streptomycin) or pyrazinamide, sometimes for more than 6 months. The extended period is necessary because of the slow-growing nature of the bacteria and to enable the drugs to permeate granulomatous and tubercle lesions.

Items 19-22

Figure 1.1 is a diagrammatic cross section of a flagellated bacterium with various structures labeled. Match the labeled structure with the most appropriate description of the structure in the items below.

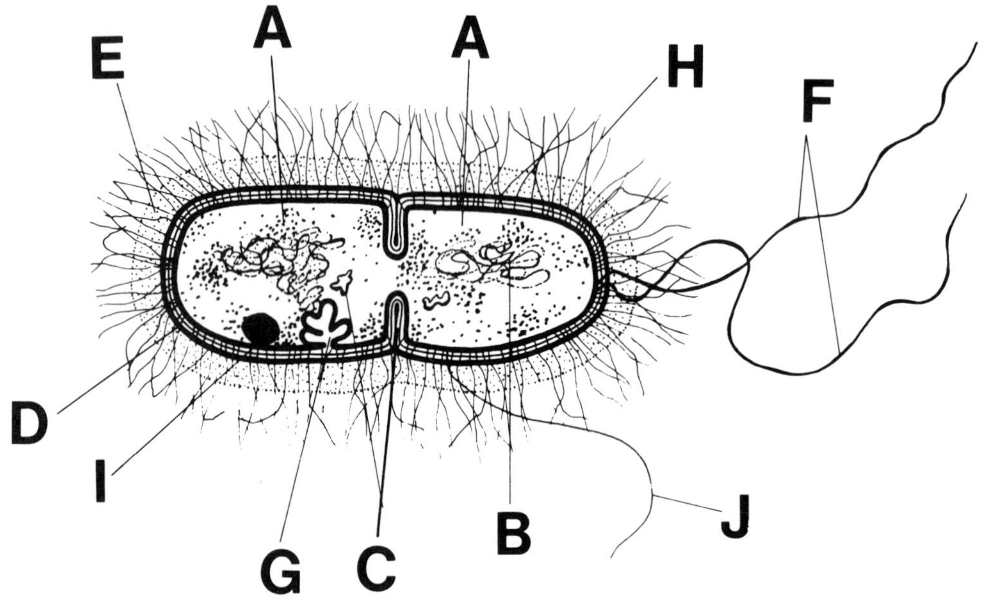

Figure 1.1

19. Genetic information organelle

20. Motility organelle

21. Antiphagocytic organelle

22. Energy storage organelle

Items 23-27

Many acute bacterial diseases follow a predictable time course with respect to the symptoms expressed in the individual. The following diagram might represent the pattern of an individual suffering from a mild bacterial infection of the throat.

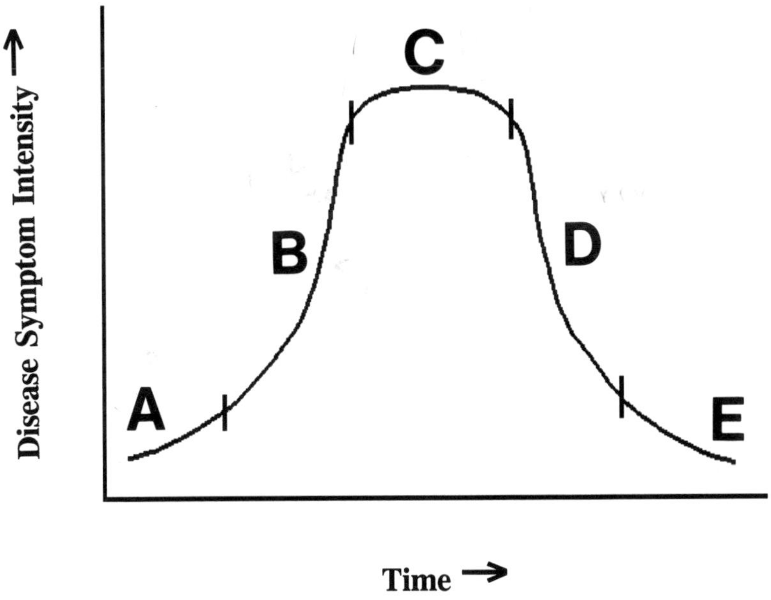

Figure 1.2

For each of the following phases, select the corresponding time location indicated in the above diagram.

23. Period of disease onset

24. Recovery to pre-disease state

25. Period of incubation

26. Prodromal period

27. At what point would the above diagram diverge or dramatically change if the plot of an immunodeficient individual were superimposed?

ANSWERS AND TUTORIAL ON ITEMS 19-22

The answers are: **19-B; 20-F; 21-E; 22-I**. Procaryotes are functionally compartmentalized, but lack the well-defined membranous boundaries found around discrete organelles in eucaryotic cells. An area of high electron density in electron micrographs corresponds to the area of high DNA concentration. This nucleoid region (B) contains the bacterial DNA and newly transcribed RNA's. Translation takes place throughout the cell cytoplasm on ribosomes (A).

Bacteria move by flagellar (F) rotation. They may be situated at one end (monotrichous) or along the entire length of the bacterium. Smaller appendages called pili are found on some bacteria. They are not motility organelles but may function in conjugation or adherence to tissues receptors.

The bacterial capsule (E), when present, consists of complex polysaccharides in almost all bacteria. It serves as a virulence factor in that it may act to prevent engulfment by phagocytic cells or mediate adhesion to tissues receptors within a host animal. The antigenic property of the bacterial capsule is sometimes useful in identifying different species and strains.

Some bacteria may harbor a variety of storage vacuoles (I) within the cytoplasm. Polymerized metaphosphate or lipid are examples of inclusions which serve to store energy reserves for the organism. The presence of different inclusions or granules can be detected using differential stains. Other structures represented within the diagram are septum formation (C), call wall (D), mesosome (G), pili (H) and sexual pili (J).

ANSWERS AND TUTORIAL ON ITEMS 23-27

The answers are: **23-B; 24-D; 25-A; 26-B; 27-A**. Most infectious diseases follow four stages consisting roughly of an incubation period (A), marked by the first encounter with the pathogen and early establishment within the animal; the period of disease onset (also called the prodromal period) (B) during which symptoms progressively develop, some appearing nonspecific; the maximum illness (acute disease) period (C) marked by "characteristic" symptoms; and finally the recovery period (also called the convalescent period) (D), marked by a decrease in symptoms as the body gradually returns to the pre-disease condition (E).

The incubation period of most diseases varies. This is due, in part, to the immunological state and overall health of the body. Infections in immunocompromised individuals would therefore tend to display shorter incubation periods since the body would lack prior cross protecting humoral and cellular immunities.

Items 28-31

A young healthy woman eagerly expected the birth of her first child. Prenatal care instructions were followed and all indications appeared normal. Unfortunately, a still-born fetus with neonatal meningitis was the outcome. The mother was an asymptomatic carrier of the pathogen.

28. Such an occurrence could result from the intrauterine or vaginal colonization by

    (A) Listeria monocytogenes
    (B) Bacteroides fragilis
    (C) Clostridium difficile
    (D) Clostridium perfringens
    (E) Gardnerella vaginalis

29. If a sample of the CSF from the infant were gram stained, observed bacteria would be

    (A) gram-negative cocci
    (B) gram-negative rods
    (C) gram-variable rods
    (D) gram-positive cocci
    (E) gram-positive rods

30. The drug of choice in the treatment of this pathogen is

    (A) doxycycline
    (B) penicillin
    (C) cephalosporin C
    (D) streptomycin
    (E) amphotericin B

31. The pregnant woman most likely was exposed to the pathogen through

    (A) the bite of an arthropod vector
    (B) contaminated blood products
    (C) inhalation of aerosol droplets
    (D) ingestion of contaminated foods
    (E) an infected pet

Items 32-35

A 60 year-old patient in the burn ward of a hospital develops an infection in the burned tissue. Initial treatment with chloramphenicol and tetracycline was unsuccessful. Cultures of the infected site revealed gram-negative rods that grew on blood agar, EMB, and trypticase soy agar. On the latter medium they produced a bluish-green pigmentation and aromatic odor. The organism was susceptible to carbenicillin, gentamicin, amikacin and polymyxin B.

32. This organism is most likely

    (A) Acinetobacter calcoaceticus
    (B) Alcaligenes faecalis
    (C) Flavobacterium sp.
    (D) Pseudomonas aeruginosa
    (E) Citrobacter sp.

33. The most probable sources of nosocomial infections include each of the following **EXCEPT**:

    (A) endogenous organisms
    (B) hospital staff
    (C) medical devices
    (D) hospital environment/surfaces
    (E) therapeutic solutions

34. A major organism associated with nosocomial infections of surgical wounds is

    (A) Staphylococcus aureus
    (B) Alcaligenes faecalis
    (C) Streptococcus pyogenes
    (D) Streptococcus pneumoniae
    (E) Bacillus cereus

35. A reasonable approach to combatting the spread of nosocomial infections within a hospital would include all of the following measures **EXCEPT**:

    (A) epidemiologic study of infection prevalence
    (B) strict quality/hygiene control measures
    (C) containment of infectious outbreaks
    (D) vigorous education programs
    (E) general administration of high antibiotic levels

ANSWERS AND TUTORIAL ON ITEMS 28-31

The answers are: **28-A; 29-E; 30-B; 31-D**. Listeria monocytogenes is a gram-positive rod and normally only causes asymptomatic transient infections in most adults. Patients infected with Listeria may become septic as the bacteria enter the blood. The bacteria may also enter the central nervous system, especially in immunocompromised adults. Pregnant women are most at risk since the organism can either infect the fetus by crossing the placental membrane or infect the neonate during transit through the birth canal. Stillbirths, abortions, premature births and neonates suffering from disseminated abscesses and/or granulomas may result.

A prime source of infection is through contaminated foods. Listeria is known to survive both pasteurization processes and refrigeration. Penicillin, ampicillin, erythromycin and chloramphenicol are all effective in treating the disease.

ANSWERS AND TUTORIAL ON ITEMS 32-35

The answers are: **32-D; 33-E; 34-A; 35-E**. Hospital-acquired, or nosocomial, infections require continued surveillance and adherence to proper hygiene practices on the part of all health care providers. Pseudomonas aeruginosa is among the most common nosocomial pathogens and most difficult to treat due to its broad range of antibiotic resistance. Burn patients are among those at high risk from this pathogen due to its ability to survive under conditions of low nutrient requirement on surfaces and stagnant water. Growth properties on EMB, colony pigmentation and the aromatic compounds they produce make them easy to identify. Contamination of hospital solutions (e.g. IV fluids or antibiotics) by Pseudomonas is less likely than floors, walls, drinking water, dialysis machinery, whirl pools and machinery. Medical personnel may harbor the bacteria and pass it, unknowingly, during routine patient examinations.

Staphylococcus aureus, a gram-positive coccus, is a major nosocomial agent infecting surgical wounds and injuries. Many individuals carry staphylococcal organisms on their skin and upper respiratory tract. It is the most prevalent of the gram-positive cocci causing hospital-acquired infections, followed by the Streptococci.

All hygienic measures such as washing, disinfection and sterilization of potential sources of contamination will reduce the probability of pathogen spread. The indiscriminate use of antibiotics, in the long run, will only result in the selection of more fastidious antibiotic-resistant microbes.

-gram + cocci
1. staph
2. strep

- gram - cocci
1. nisseria (diplococci)

- gram + rod
1. clostridium
2. cornybacterium
3. listeria
4. bacilus

- gram - rod
- everything else

mycoplasma : non-staining
- Ureaplasma
(no cell wall)

Items 36-39

A 34 year-old male homosexual closely adhered to a personal policy of annual physical checkups and routinely received a clean bill of health. Three weeks after returning from a business trip to California, he noticed a small circular lesion on tip of his penis. The area became more sensitive and began to ooze fluid. Medical examination revealed lymphocytes and plasma cells and no unexpected gram-stainable bacteria. A serum sample showed a positive VDRL test with a titer of > 1:64.

36. A presumptive diagnosis would be

    (A) nongonococcal urethritis
    (B) primary syphilis
    (C) secondary syphilis
    (D) tertiary syphilis
    (E) latent syphilis

37. A confirmatory diagnostic test would include a/an

    (A) FTA-ABS test
    (B) RPR test
    (C) darkfield microscopic test
    (D) isolation on enriched nutrient media
    (E) Western blot

38. If the patient does not receive medical treatment, and the lesion clears, there remains the possibility of

    (A) nongonococcal urethritis
    (B) primary syphilis
    (C) secondary syphilis
    (D) tertiary syphilis
    (E) latent syphilis

39. The bacteria responsible for the disease will

    (A) remain confined to the chancre and remain infectious
    (B) remain confined to the chancre and become avirulent
    (C) not affect regional lymph nodes
    (D) disseminate to regional lymph nodes and infect other organs
    (E) produce many similar lesions on the penis

ANSWERS AND TUTORIAL ON ITEMS 36-39

The answers are: **36-B; 37-A; 38-C; 39-D**. This man has primary syphilis caused by the bacterium Treponema pallidum. The early disease is marked, in most cases, by a genital or anal lesion called a chancre. Since the bacteria do not stain readily in the Gram stain, their presence, in microscopic smears taken from the lesion, may go unnoticed. A humoral response is usually detectable after two weeks of exposure. It is marked by the presence of nonspecific (reagin) antibodies, detectable using the classical VDRL (Venereal Disease Research Laboratory) test or RPR (rapid plasma reagin) test. A confirmatory diagnostic test would include the FTA-ABS (fluorescent treponemal antibody-absorbed test) which uses antigen prepared from the bacteria. Darkfield microscopic examination, though useful in visualizing the spirochetes is not species-specific. The bacteria do not grow on nutrient media. Western blot analysis, though feasible, has not been applied to the routine identification of the treponemal organisms.

Primary syphilis, if untreated, will progress to an advanced stage, secondary syphilis, marked by a dissemination of the organism to regional lymph nodes and other organs. A maculopapular rash or moist papules on the skin may develop.

Any sexual contacts of this man may also harbor the pathogen and display similar lesions in areas of sexual contact. All people who come in sexual contact with this man should receive medical treatment and counseling for the prevention of sexually transmitted diseases.

Items 40-44

A young undergraduate college student, majoring in botany, went on a field trip with her classmates to search the Delaware shoreline for representative plants. The area included both shoreline and a deciduous forest inhabited by small rodents. The students entered both areas. A week following the trip she experienced flu-like symptoms which included high fever, headaches, muscle aches, general malaise and a stiff neck. A large 3-4 cm circular red blotch developed on her lower leg and became sensitive to the touch and gradually increased in size.

40. On the basis of the history, the symptoms closely fit a preliminary diagnosis of

(A) Weil's disease
(B) leptospirosis
(C) Lyme disease
(D) epidemic relapsing fever
(E) plague

41. The disease is transmitted by the bite of the

(A) louse
(B) gnat
(C) deer tick
(D) mosquito
(E) wood tick

42. The causative agent of the disease is

(A) Borrelia burgdorferi
(B) Leptospira interrogans
(C) Leptospira icterohemarrhagiae
(D) Borrelia recurrentis
(E) Rickettsia rickettsii

43. The untreated disease will usually resolve itself, however several months following the initial infection, a small percentage of infected individuals will experience a "second stage" of infection involving

(A) visual damage
(B) renal failure
(C) neurological abnormalities
(D) disseminated skin lesions
(E) synovial infection and arthritis

44. A third stage of the disease, occurring months to years after the initial infection, may occur involving

(A) visual damage
(B) renal failure
(C) neurological abnormalities
(D) disseminated skin lesions
(E) autoimmune complexes and arthritis

ANSWERS AND TUTORIAL ON ITEMS 40-44

The answers are: **40-C; 41-C; 42-A; 43-C; 44-E**. Lyme disease is caused by the bacteria Borrelia burgdorferi. The pathogen is transmitted through the bite of an arthropod vector, namely the deer tick Ixodes dammini and I. pacificus. The ticks are very small, measuring less than 2 mm, rendering them easily overlooked when they attach and feed. The ticks have been found on other animals including mice, rats, skunks, opossums and dogs.

The primary lesion grows into an enlarged rash as the spirochete proliferates and invades the blood and lymphatics. The lesion usually disappears after several weeks. Some individuals progress into a second stage of infection involving neurologic or cardiac abnormalities. An arthritic response may develop months to years after the initial infection, probably due to an autoimmune response. This is indicative of the third stage of Lyme disease.

Items 45-48

A 65-year old woman, previously in excellent health, experienced a moderate intestinal tract infection. She had attended a neighborhood oriental restaurant at lunchtime. Within five hours she developed nausea and vomiting. Food poisoning was an obvious possibility. Samples of the different dishes consumed by the woman all contained low levels of different bacteria. A rice dish showed a high concentration of aerobic gram-positive rods.

45. The appropriate treatment for the woman should consist of

(A) only symptomatic aid
(B) antibiotics effective against gram-positive bacteria
(C) antibiotics effective against gram-negative bacteria
(D) intravenous drug therapy
(E) administration of an antitoxin

46. A possible causative organism would be

(A) Staphylococcus aureus
(B) Bacillus anthracis
(C) Clostridium botulinum
(D) Bacillus cereus
(E) Clostridium difficile

47. The symptoms experienced by the patient are most probably due to

(A) toxin production
(B) bacteremia
(C) sporulation
(D) massive intestinal bacterial growth
(E) an inflammatory response

48. All of the following are gram-positive bacilli **EXCEPT**:

(A) Bacillus sp.
(B) Clostridium sp.
(C) Corynebacterium sp.
(D) Listeria sp.
(E) Campylobacter sp.

Items 49-52

A migrant farm worker disregarded a tender area in his mouth. It eventually developed into a enlarged abscess requiring medical treatment. Fluid obtained from the abscess revealed numerous gram-negative rods. Nothing grew on a streaked blood agar plate incubated at 37°C.

49. An appropriate next step would be to

(A) repeat the blood agar culture
(B) incubate the culture at different temperatures
(C) streak and incubate the culture under anaerobic conditions
(D) perform a spore stain
(E) perform an acid-fast stain

50. Which of the following is most likely responsible for the abscess?

(A) Bacteroides fragilis
(B) Bacteroides corrodens
(C) Bacteroides melaninogenicus
(D) Escherichia coli
(E) Pseudomonas aeruginosa

51. Oral infections are usually initiated by

(A) trauma to the mucosa
(B) superinfection by another microbe
(C) adverse chemotherapy treatment
(D) hereditary factors
(E) age-dependent factors

52. Which of the following is the most effective drug of choice used in the treatment of the causative agent?

(A) kanamycin
(B) vancomycin
(C) penicillin
(D) cephalosporins
(E) clindamycin

ANSWERS AND TUTORIAL ON ITEMS 45-48

The answers are: **45-A; 46-D; 47-A; 48-E**. The spores of gram-positive aerobes such as the Bacillus sp. can cause gastrointestinal illnesses which mimic food poisoning caused by other organisms. Grains may harbor spores not destroyed by routine cooking and reheating. This allows the vegetative bacteria to colonize and release toxins into prepared foods. This is most true of the aerobic bacilli such as Bacillus cereus.

The gastroenteritis usually lasts briefly without the development of bacteremia. Transient colonization of the gastrointestinal tract occurs without the necessity of any specific antibiotic treatment. Antidiarrhetics are usually sufficient remedy for the infection which will resolve itself.

ANSWERS AND TUTORIAL ON ITEMS 49-52

The answers are: **49-C; 50-C; 51-A; 52-E**. Bacteroides sp. are gram-negative anaerobic organisms commonly associated with abscesses. They are normally found in the body, but are opportunistic, taking advantage of breaks in the mucosal lining. Bacteroides melaninogenicus causes oral, pharyngeal and pulmonary abscesses.

Standard blood agar cultivation will not yield colony growth. Anaerobic conditions are necessary with the inclusion of antibiotics, such as kanamycin and vancomycin present in the culture media, to reduce contamination by other bacteria.

Bacteroides melaninogenicus infections respond favorably to treatment with clindamycin. Penicillin sensitivity varies with different species of Bacteroides; B. fragilis is resistant, whereas B. melaninogenicus is sensitive.

Items 53-57

A concerned mother took her 10 month-old daughter to the pediatrician after noticing bloody diarrhea, flatulence and a slight fever of 101°C in her child. Bacterial cultures revealed the presence of Salmonella sp. organisms. Other family members exhibited neither similar symptoms nor harbored a similar type of bacteria. The family had no pets; however, three weeks earlier her 10 year-old son brought home a pet turtle from his elementary school science lab. Cultures taken from the turtle also grew up Salmonella. The mother helped her son care for the class pet for two weeks.

53. The daughter is most likely suffering from

(A) septic shock
(B) bacteremia
(C) enterocolitis
(D) typhoid
(E) septicemia

54. The illness is marked by the

(A) ingestion of a high ($>10^5$) dose of organisms
(B) invasion of the blood
(C) repeated fevers
(D) CNS involvement
(E) high fatality rate

55. The most common cause of the above illness is

(A) Salmonella choleraesuis
(B) Salmonella typhimurium
(C) Salmonella typhi
(D) Salmonella paratyphi
(E) Salmonella enteritidis

56. The most likely source of infant exposure to the pathogen was through

(A) inhalation of aerosol droplets from the mother
(B) inhalation of aerosol droplets from the son
(C) direct contact with the contaminated turtle
(D) food contaminated by the mother
(E) skin abrasions

57. Medical treatment for the daughter would consist of

(A) nothing since such infections self-resolve
(B) supportive treatment only
(C) electrolyte replacement
(D) antibiotics and electrolyte replacement
(E) antitoxin and antibiotics

ANSWERS AND TUTORIAL ON ITEMS 53-57

The answers are: **53-C; 54-A; 55-B; 56-D; 57-D**. Salmonella infections are associated with the ingestion of relatively large infecting doses of organisms ($>10^5$) resulting in enterocolitis. The disease in characterized by bacterial invasion of the mucosa of the small and large intestines, resulting in inflammation and diarrhea. Salmonella typhimurium is the primary cause of enterocolitis even though all members of the genus have been implicated. More invasive infections can lead to typhoid or septicemia.

Enterocolitis epidemics have been traced to sources such as pet reptiles and poultry as primary sources which subsequently contaminate food. Adequate refrigeration, proper hygiene and thorough cooking will restrict pathogen growth, reduce the levels of contamination below dangerous levels or inactivate toxin accumulation, respectively.

The transmission of Salmonella from household pets, such as reptiles and birds, to members of a family has been documented. Direct contact between a family member and the pet is not necessary since the bacteria may remain viable in contaminated water, dried feces, feathers, cages and pet foods. Infants are usually infected by contact with family members who have recently handled the pet. Low dose exposure may explain the long incubation interval and absence of infection in older individuals who may have previous exposure to Salmonella sp. and therefore some resistance.

Antibiotic treatment in not normally necessary since infections are self-limiting; however, ampicillin or chloramphenicol treatment is indicated in the cases involving young children or people with chronic disease.

Items 58-61

A small clinic located in a Texas boarder town receives a high volume of patients suffering from a variety of sexually transmitted diseases. Many of the patients are foreign workers lacking residency documentation and therefore follow up treatment is difficult. Local prostitutes serve both the foreign workers and the local residents, thus providing a critical link for the introduction of a variety of diseases previously not encountered. One male patient complained of a venereal infection which was a 1 cm, irritated, soft chancre-like, penile lesion. He also had regional lymphadenitis.

58. A gram-negative rod isolated from both the lesion and pus taken from the lymph node could represent an infection caused by

(A) Haemophilus ducreyi
(B) Neisseria gonorrhoeae
(C) Treponema pallidum
(D) Chlamydia trachomatis
(E) Treponema carateum

59. The disease is

(A) yaws
(B) chancroid
(C) pinta
(D) lymphogranuloma venereum
(E) gonorrhea

60. Diagnosis is based primarily on

(A) Giemsa stain
(B) Tzanck smear
(C) darkfield microscopy
(D) culture isolation on chocolate agar supplemented with growth factors
(E) serum antibody titer

61. Which of the following is the most effective drug of choice used in the treatment of the causative agent?

(A) kanamycin
(B) vancomycin
(C) penicillin
(D) cephalosporins
(E) clindamycin

Items 62-65

In August a 27 year-old man with left-sided facial pain visited the emergency room. He was unable to open his mouth because of facial muscle spasms and had been unable to eat for 3 days because of the pain and stiffness. A clinical history revealed that the young man had been searching for scrap metal in a junk yard and accidentally stepped of a jagged piece of metal projecting from the soil. He self-treated the wound and ignored it until he started experiencing the facial tightness and pain.

62. The symptoms experienced by the young man are consistent with a presumptive diagnosis of

(A) gangrene
(B) tetanus
(C) carbuncles
(D) paronychia
(E) furuncles

63. All of the following would be recommended in the treatment of the patient **EXCEPT**:

(A) tetanus toxoid
(B) tetanus immune globulin
(C) penicillin (i.v.)
(D) steroids
(E) thorough wound debridement

64. The causative agent, <u>Clostridium</u> <u>tetani</u>, elicits the disease by the production of

(A) tetanospasmin
(B) protein endotoxin
(C) cell wall-bound toxin
(D) capsule carbohydrate
(E) lipopolysaccharide

65. The toxin produced by <u>Clostridium</u> <u>tetani</u>

(A) is produced in excess by the multiplying bacteria
(B) is produced in minute amounts
(C) resists proteolysis
(D) is heat stable
(E) induces a strong immune response in infected individuals

ANSWERS AND TUTORIAL ON ITEMS 58-61

The answers are: **58-A; 59-B; 60-D; 61-D**. Chancroid is a relatively rare venereal disease limited primarily to North America. It is caused by the bacteria Haemophilus ducreyi. The infection is acquired through direct sexual contact. A non-indurated chancre, also referred to as a soft chancre, which distinguishes it from the indurated hard chancre associated with syphilis, develops on the penis. The bacteria may cause additional genital lesions along with invasion of the lymph nodes. An inflammatory response results in the recruitment of lymphocytes and plasma cells by relatively few bacteria.

Diagnosis in based on clinical symptoms and the isolation of H. ducreyi on chocolate agar, supplemented with hematin (X factor). Like some members of the genus Haemophilus, it has a natural requirement for hematin for growth. Other species such as H. influenzae require both heme and NAD (V factor) for growth. Competing bacteria are reduced during isolation attempts by the inclusion of vancomycin in the growth agar to reduce overgrowth by other organisms.

Antibiotic resistance is a growing concern with Haemophilus sp. Many have acquired penicillin resistance through the acquisition of plasmid-derived β-lactamases. Penicillin is therefore prescribed less frequently. The new cephalosporins or sulfonamides are drugs of choice. Combination drug therapy with erythromycin and sulfonamide is also effective.

ANSWERS AND TUTORIAL ON ITEMS 62-65

The answers are: **62-B; 63-D; 64-A; 65-B**. Clostridium tetani is an anaerobic, soil-borne bacterium which can cause tetanus. The spores of the bacteria produce a powerful exotoxin called tetanospasmin. The growth of the bacteria in the anaerobic conditions of the deep wound leads to spore and endotoxin formation. Victims are usually older than 20 years due to a diminution of vaccine antibody titer. Most individuals have therefore been immunized with the traditional diphtheria-pertussis-tetanus (DPT) vaccine, but full immunity is relatively short lived, even after a recent booster, in some individuals. Antitoxin level may last for 10 years in properly vaccinated individuals, with boosters at 10-year intervals advised to maintain protective titers.

Wounds, leading to tetanus, are most prevalent during the summer months. Complete wound cleaning, along with tetanus toxoid, immune globulin (anti-toxin) and antibiotic (such as penicillin) are recommended in instances where disease develops.

Tetanospasmin is a potent extracellular protein neurotoxin released by cellular autolysis. It is produced in minute amounts by the small number of wound bacteria. The toxin is heat-labile, and sensitive to enteric proteolytic enzymes.

Items 66-69

Five members of a family (age range 19-55) had onset of fever (four family members), anorexia (three), sore throat (four), cough (three), headache (three), vomiting (two) and myalgia (three). Evaluation of three patients included chest X-rays. All three showed diffuse interstitial pulmonary infiltrates. Nasal aspirates revealed no unusual findings for either culturable bacteria or viruses. Family pets included a small terrier, four years in the household, and a cockatiel, purchased a month prior to the onset of symptoms.

66. The pneumonia exhibited by several family members supports a presumptive diagnosis of

(A) influenza
(B) pneumococcal pneumonia
(C) respiratory syncytial virus
(D) psittacosis
(E) primary atypical pneumonia

67. Adult family members should respond to treatment with

(A) penicillin
(B) amantadine
(C) doxycycline
(D) interferon
(E) amoxicillin

68. The source of the infection is most likely

(A) the family dog
(B) the family cockatiel
(C) through normal daily contact outside the home
(D) contaminated foods
(E) contaminated drinking water

69. A rapid and reliable diagnostic test used to establish chlamydial infections is the

(A) cell culture isolation test
(B) complement fixation test
(C) Frei (skin) test
(D) enzyme-linked immunosorbent assay (ELISA) test
(E) Giemsa stain

ANSWERS AND TUTORIAL ON ITEMS 66-69

The answers are: **66-D; 67-C; 68-B; 69-D**. The family members have psittacosis due to an infection by Chlamydia psittaci transmitted from the cockatiel. The CDC reports only 100-250 cases of psittacosis each year in the United States. Symptoms may vary even within a focal point from mild symptomatic to more pronounced illness. Usually an encounter with pet birds or avian farm animals can be linked to the onset of the disease. Importers and distributors of pet birds are required to maintain animals in quarantine for a week before shipment to distributors. Chlortetracycline placed in animal feed, for a 45 day treatment period, may prevent the spread of chlamydia among expensive bird stocks.

Examination of euthanized animals will usually show the presence of chlamydial antigen from throat and cloacal swabs. Serum antibody against chlamydia and hepatosplenomegaly may develop in advanced cases.

Chlamydial infections in adults respond favorably to the tetracyclines, when administered over a 2-3 week period. A new antibiotic, azithromycin, has been shown effective as a single-dose treatment.

The rapid (3 hours), sensitive and automated ELISA for the presence of high serum antichlamydial antibody has become widely used. It is nearly as sensitive as the classical cell culture isolation technique. Unfortunately, the latter is both labor intensive and time-consuming (4-6 days).

Items 70-73

A group of 25 businessmen celebrated at a New Year's Day party. In addition to alcoholic refreshments and a wide assortment of cooked foods, the feast included fresh oysters and clams. On January 2, twenty-three of them began to experience gastrointestinal illnesses which included diarrhea (86%), nausea (82%), abdominal cramps (65%), fever (56%) and vomiting (35%). In most cases the illnesses lasted 15-17 hours. Only the mollusks were consumed by all of the affected individuals. Bacteriological cultures taken from the mollusks produced significant colony numbers of gram-negative rods including Escherichia coli, Vibrio parahaemolyticus, and Clostridium perfringens.

70. Numerous enteric pathogens are known to be transmitted through the consumption of raw shellfish. All of the following can be transmitted by consumption of raw shellfish **EXCEPT**:

(A) Salmonella
(B) Shigella
(C) Campylobacter
(D) hepatitis A virus
(E) Serratia

71. A low molecular weight protein enterotoxin, comprising part of a bacterial spore is produced by

(A) Clostridium perfringens
(B) Escherichia coli
(C) Campylobacter jejuni
(D) Shigella dysenteriae
(E) Salmonella typhi

72. The halophilic nature and tendency to induce diarrhea, following colonization of the intestinal tract, makes which of the following organisms a primary candidate for causing the above illness?

(A) Clostridium perfringens
(B) Vibrio parahaemolyticus
(C) Campylobacter jejuni
(D) Bacillus cereus
(E) Salmonella typhi

73. High levels of Escherichia coli recovered from shellfish samples is indicative of

(A) normal environmental conditions
(B) improper storage of the shellfish
(C) sewerage pollution of shellfish beds
(D) shellfish eating patterns
(E) poor sample collecting procedures

ANSWERS AND TUTORIAL ON ITEMS 70-73

The answers are: **70-E; 71-A; 72-B; 73-C**. Public Health Departments in most states closely regulate the sale of shellfish. Some guidelines include the tagging and tracing of such foods especially those crossing interstate lines. Gastroenteritis is not always traced to a specific causative agent, but rather several potential pathogens.

A variety of bacterial, viral and parasitic pathogens are associated with the consumption of raw shellfish. Serratia sp., are not normally associated with the high salt conditions of shellfish-associated diseases, but rather nosocomial infections.

Clostridium perfringens is a spore forming bacterium listed in item 71. The enterotoxin is associated with the spore wall. It is released during germination.

Many Vibrio organisms are water-borne. They have a high tolerance to sea water, thus they are commonly isolated from shellfish and fish. Disease results from colonization of the bacteria in the intestines and release of exotoxins. Diarrheal infections caused by Vibrio parahemolyticus respond to antibiotic therapy which also diminishes the likelihood of secondary infections and shortens the duration of diarrhea. Tetracycline, ampicillin and chloramphenicol are among the drugs used for this purpose.

The abruptness of the symptoms following consumption of the foods imply the presence, in the foods, of significant toxin. However, rapid colonization of the gastrointestinal tract and toxin production are characteristics of certain bacteria such as V. parahemolyticus. Improper storage and extended shipping intervals may provide additional favorable conditions for bacterial growth and toxin production. High colony counts of Escherichia coli are indicative of fecal contamination. The waters of shellfish collection may have been polluted with human sewage.

Items 74-77

A female taxidermist for the Smithsonian Institution was on an assignment in Brazil. Her work involved the processing of small mammal skins for transport to her home laboratory in Washington, D.C. During the second month of her work in the jungles, she experienced an acute illness consisting of severe headache, chills, fever, sweating, loss of appetite and muscle spasm in the lower back and hip. Ampicillin administered by a local physician did not relieve the symptoms. A week later her condition worsened and she returned to Washington and was examined at The George Washington University Medical Center. Her condition had progressed to include a pain in her right axilla, surrounding edema and a dry cough. A presumptive diagnosis of bubonic plague was made.

74. Aspiration of an affected lymph node would show the presence of

(A) gram-positive rods
(B) gram-negative bipolar rods
(C) gram-positive cocci in pairs
(D) gram-negative diplococci
(E) gram-variable spirochetes

75. Bacteriological cultivation of the lymph node aspirate would most likely yield colonies of

(A) Yersinia enterocolitica
(B) Yersinia pseudotuberculosis
(C) Salmonella typhi
(D) Yersinia pestis
(E) Shigella dysenteriae

76. An appropriate therapy for the treatment of the disease would consist of

(A) 1-day on streptomycin
(B) 10-days on streptomycin
(C) 20-days on streptomycin
(D) 30-days on streptomycin
(E) 40-days on streptomycin

77. The researcher probably encountered the bacterium through the bite of an infected

(A) tick
(B) flea
(C) mosquito
(D) louse
(E) fly

ANSWERS AND TUTORIAL ON ITEMS 74-77

The answers are: **74-B; 75-D; 76-B; 77-B**. Bubonic plague is caused by the gram-negative bacteria *Yersinia pestis*. The bacterial infection is acquired through the bite of an infected flea followed by rapid multiplication in regional lymph nodes and dissemination to other organs. In this case, the right axilla was affected. Localized necrosis results from tissue damage and toxin production. Recovery confers natural immunity against reinfection. The more deadlier pneumonic plague, involving person-to-person transmission, may result from the inhalation of the bacteria.

A primary task of the project was the collection of small mammal pelts. Many such animals harbor parasites including fleas and ticks. The researcher, in an effort to economize on materials required for the study, substituted the usual chloroform used to euthanize animals, for less bulky nembutal. The latter is injected into animals; however, the former kills both the small animals and their harbored fleas. Therefore live fleas could have easily come into contact with the researcher and transmitted the pathogen.

Yersinia pestis has bipolar staining patterns and is easily recovered from bubo aspirate, blood and sputum. Vaccination against plague should be boosted every 1-2 years for those traveling in endemic regions. Treatment with streptomycin for 10 days will usually cure bubonic plague.

Items 78-81

The Enterobacteriaceae produce a variety toxins and virulence factors with different modes of action. For each of the following toxins, select the most likely description in the items below.

(A) LPS endotoxin
(B) enterotoxin LT and/ or ST
(C) Shiga toxin
(D) Vi polysaccharide
(E) pili

78. Surface antigen found on *Salmonella typhi* that enhances invasion of host cells.

79. Potent endotoxin component found in the cell wall of all Enterobacteriaceae.

80. Potent exotoxin which induces eucaryotic cell death by blocking protein synthesis.

81. *Escherichia coli* toxin which induces fluid release from the intestines.

ANSWERS AND TUTORIAL ON ITEMS 78-81

The answers are: **78-D; 79-A; 80-C; 81-B**. The surface of Salmonella typhi contains a virulence (Vi) polysaccharide (D) which comprises part of the total surface. Presence of this antigen enables the bacteria to invade and multiply within cells of the small intestine.

The lipid A moiety of lipopolysaccharide (LPS) (A) is a highly toxic substance (endotoxin) found in the cell wall of all Enterobacteriaceae and other gram-negative bacteria. The polysaccharide core associated with the O-antigen side chains forms the basis of many antigenic differences found on bacteria. The LPS forms a barrier from the external environment. The highly charged hydrophobic nature of the LPS also helps to maintain hydrophilic molecules within the cell.

Enterogenic Escherichia coli produce two toxins; LT heat-labile and ST heat-stable (B). Both are encoded by a small plasmid. The LT and ST toxins are similar in action to that of the cholera toxin, causing a stimulation of adenylate cyclase and guanylate cyclase, respectively. The net effect is the dramatic release of fluids from the affected cells.

The Shiga toxin (C) is a potent exotoxin. It is known to have enterotoxic, neurotoxic and cytotoxic properties. The collective activity is a property of a complex consisting of two different molecules (A/B). B is involved with binding properties to host receptors. A eventually binds to host cell ribosomes, causing their inactivation. Both subunits of the toxin are encoded on the bacterial chromosome.

Bacterial pili (E) are thin straight projections from the bacterial cell surface. They are shorter, straighter and thinner than flagella. They are made up of aggregates of pilin monomers. They enable the bacteria to adhere to specific surfaces and therefore permit colony formation. They are also used in bacterial conjugation.

Items 82-87

An industrious young man took a summer job working with a craftsman who used natural fibers in the creation of custom garments and crafts. The job included the combing and dying of different wools for processing. Two weeks after working with a batch of imported wool the young man developed an abrupt fever, chills and a dry hacking cough. His physician noted a patchy interstitial pneumonia and enlarged spleen. No skin rash developed. Tetracycline was prescribed and the symptoms quickly disappeared after 7 days. Paired serum samples were taken during the acute illness period and three weeks following complete recovery.

82. A Weil-Felix test on the convalescent serum sample would most likely yield the following pattern:

	OX-19	OX-2	OX-K
(A)	+	+	+
(B)	+	+	-
(C)	+	-	-
(D)	-	-	-
(E)	-	-	+

83. The antigen used in the Weil-Felix test are derived from

(A) Shigella
(B) Salmonella
(C) Pseudomonas
(D) Proteus
(E) Rickettsia

84. The disease most likely experienced by the young man was

(A) Rocky Mountain spotted fever
(B) Murine typhus
(C) Brill's disease
(D) Rickettsialpox
(E) Q fever

85. The causative agent is

(A) Rickettsia rickettsii
(B) Rickettsia typhi
(C) Rickettsia prowezekii
(D) Rickettsia akari
(E) Coxiella burnetii

86. Transmission of the disease is through

(A) the bite of a flea
(B) the bite of a tick
(C) the bite of a louse
(D) the bite of a mite
(E) inhalation of agent

87. Final diagnosis would be dependent of

(A) cultivation of synthetic media
(B) electron microscopic observation
(C) immunofluorescence using species-specific antibody
(D) eventual rash development
(E) sugar fermentation results

ANSWERS AND TUTORIAL ON ITEMS 82-87

The answers are: **82-D; 83-D; 84-E; 85-E; 86-E; 87-C**. The causative agent of Q fever, Coxiella burnetii, is a rickettsial organism which does not require an arthropod vector for transmission. Infections occur through inhalation of air-borne, dried Coxiella from contaminated wools and animal hides. The dehydrated bacteria remain viable in the dried state for several years awaiting the suitable environment of a human or animal upper respiratory tract.

The rickettsial disease Q fever differs from other rickettsial infections by several unique features. The traditional indirect agglutination screening test for rickettsial infections, the Weil-Felix test, is negative. The test relies on the agglutination of cross-reacting rickettsial antibodies against three standard Proteus vulgaris strains. The test is presumptive in that some rickettsial diseases yield similar agglutination patterns and still others, such a Q fever, yield no agglutinations. Immunofluorescence or complement fixation tests, using species-specific antibody, would identify exposure to Coxiella.

Q fever, unlike many other rickettsial infections, does not result in the expression of a characteristic rash pattern. The result is a greater difficulty in identifying the disease. Q fever, along with Rocky Mountain spotted fever occurs more frequently in the U.S. than other rickettsial infections.

Items 88-93

A second year medical student from the George Washington University Medical School chose to spend the summer working with a private relief organization. She was sent to a very rural Indian reservation in the western Badlands. Most of the visitors to the small clinic had not received periodic medical examinations. One child entered with symptoms of a tonsillitis; malaise, sore throat and fever. The tonsils and soft palate were covered by a gray-white membrane.

88. A Gram stain of throat smears would show

(A) large gram-negative rods
(B) small gram-positive rods
(C) large gram-positive cocci
(D) small gram-negative cocci
(E) small club-shaped gram-negative rods

89. The student noted the unusual feature of the bacteria and tonsils and sent throat swabs away for microbial analysis which isolated

(A) β-hemolytic streptococci on blood agar
(B) complete hemolytic golden cocci on blood agar
(C) black colonies on Tinsdale agar
(D) blue-green colonies on blood agar
(E) minute clear colonies on Eaton's agar

90. The child has

(A) a simple strep throat
(B) a staph infection
(C) diphtheria
(D) a Pseudomonas infection
(E) mycoplasma pneumonia

91. The causative agent is

(A) Streptococcus pyogenes
(B) Staphylococcus aureus
(C) Corynebacterium diphtheriae
(D) Pseudomonas aeruginosa
(E) Mycoplasma pneumoniae

92. The appearance of the membrane over the tonsils is due to

(A) the destruction of epithelial cells by toxin
(B) stimulated growth of epithelial cells
(C) overgrowth by oral bacterial flora
(D) restricted blood flow to the affected area
(E) red cell hemolysis

93. Treatment would involve

(A) vaccination with the DPT vaccine
(B) therapy with antibiotics
(C) treatment of symptoms
(D) use of antitoxin (immune globulin) and antibiotics
(E) surgery

ANSWERS AND TUTORIAL ON ITEMS 88-93

The answers are: **88-E; 89-C; 90-C; 91-C; 92-A; 93-D**. Most children are immunized against Corynebacterium diphtheriae, a club-shaped gram-negative bacterium that causes diphtheria. In isolated populations, where immunization practices are not stringently applied, children are susceptible to rapidly-spreading outbreaks. The tonsils of infected children are usually covered by a gray-white membrane consisting of bacteria, dead epithelial cells, fibrin and white cells resulting from the actions of the potent diphtheria toxin. The bacteria, Corynebacterium diphtheriae, display a characteristic appearance of club-shaped gram-negative rods.

A major priority in the treatment of diphtheria is the neutralization of the toxin by antitoxin through passive immunization and inactivation of bacteria by antibiotics. The ready-made immune globulin inactivates unbound toxin while antibiotics such as erythromycin, penicillins, cephalosporins or tetracycline kill the bacteria.

The diphtheria toxin, produced by toxigenic strains, consists of plasmid-encoded gene products composed of two subunits. The B-subunit is required for adherence to susceptible host cells. The A-subunit enters the cell and disrupts transcription by inactivating the transcriptional elongation factor 2 (EF2). The inhibition is irreversible, but results in cell death. Immunity is established by administering vaccine consisting of toxoid and booster injections.

Items 94-98

A young woman gave birth to a healthy girl. The mother came from a rural area, was in good health and had received prenatal care. At home the energetic infant scratched herself along the left thigh resulting in a slight reddening which the mother noted and treated with an antibacterial ointment. A week later the skin on the infant's leg and hands started to peel. The infant was taken to a clinic where her infection was examined, cultures were taken and treatment was initiated.

94. The exfoliation of large areas of skin from different areas of the body suggests

(A) a systemic viral infection
(B) scalded skin syndrome (Ritter's disease)
(C) toxic shock syndrome
(D) a fungal infection
(E) parasitic infection

95. Bacterial cultures taken from wounds or abscesses on the infant will most likely show the presence of

(A) _Streptococcus pyogenes_
(B) _Pseudomonas aeruginosa_
(C) _Proteus vulgaris_
(D) _Staphylococcus aureus_
(E) _Staphylococcus epidermidis_

96. The exfoliation associated with the infection is caused by

(A) α-toxin
(B) leukocidin
(C) toxic shock syndrome toxin-1
(D) coagulase
(E) exfoliative toxin

97. Strain differences of the causative agent may be distinguished by

(A) carbohydrate fermentations
(B) DNA fingerprints
(C) colony morphology of differential media
(D) animal lethal dose titers
(E) bacteriophage typing

98. Treatment of the infection usually requires

(A) drainage of any abscess and treatment with penicillin
(B) treatment with penicillin
(C) treatment with nafcillin or cloxacillin
(D) drainage of any abscess and treatment with nafcillin or cloxacillin
(E) drainage of the abscess

ANSWERS AND TUTORIAL ON ITEMS 94-98

The answers are: **94-B; 95-D; 96-E; 97-E; 98-D**. Scalded skin syndrome or Ritter's disease is one of many diseases caused by Staphylococcus aureus. The organisms are gram-positive cocci which appear randomly scattered in stained smear preparations. They colonize different portions of the skin and take opportunistic advantage of access into the body through cuts, wounds or confined locations such as certain vaginal tampons. Staphylococci in abscesses produce a variety of exotoxins which have profound pathological actions in distal parts of the body.

Exfoliative toxin causes the exfoliation of the epidermal skin layer commonly seen in scalded skin syndrome. The bacterial infection is commonly acquired by the infant, from the mother, during birth. The infant experiences a minor cut allowing the bacteria to establish an abscess. The toxin is absorbed into the blood stream and acts on the skin far from the abscess. Staphylococcus aureus produce many other toxins such as enterotoxin, which induces diarrhea and vomiting; toxic shock syndrome toxin-1, which can cause muscle cramps, fever and diarrhea; and α-toxin, which may cause skin necrosis and hemolysis.

Antigenic differences in the surface receptors are used to distinguish the different strains of staphylococci. Phage typing is based on the presence of such receptors demonstrated by the ability of different phage to attach, infect and lyse specific Staphylococcus aureus isolates.

Treatment of staphylococcal infections usually involves drainage of any prominent abscesses and antibiotic therapy. Many Staphylococcus aureus isolates produce penicillinase and are therefore resistant to penicillin. β-lactamase-resistant penicillins such as nafcillin, methicillin or cloxacillin have been effective.

Items 99-102

Two weeks prior to the Christmas vacation holiday a 6th grade student developed a severe pharyngitis. He also complained of fever and tenderness in his joints. A thorough examination of the child revealed that an infection had existed even earlier but was not reported by the child. Pain was experienced in multiple joints. His throat was grossly inflamed and throat swab culture produced small, clear, β-hemolytic colonies on blood agar.

99. The symptoms are consistent with early stages of

(A) croup
(B) scarlet fever
(C) rheumatic fever
(D) typhoid fever
(E) Q fever

100. The causative organism is

(A) parainfluenza viruses
(B) <u>Salmonella</u> <u>typhi</u>
(C) <u>Streptococcus</u> <u>pyogenes</u>
(D) <u>Coxiella</u> <u>burnetii</u>
(E) <u>Staphylococcus</u> <u>aureus</u>

101. Which of the following antigens produced by the organism is an antiphagocytic virulence factor?

(A) α-hemolysin
(B) β-hemolysin
(C) C carbohydrates
(D) M protein
(E) coagulase

102. An antibiotic effective in the treatment of the infection is

(A) penicillin
(B) tetracycline
(C) chloramphenicol
(D) sulfadiazine
(E) neomycin

ANSWERS AND TUTORIAL ON ITEMS 99-102

The answers are: **99-C; 100-C; 101-D; 102-A**. Rheumatic fever develops as a result of a prior pharyngitis caused by Streptococcus pyogenes. The organism causes an initial sore throat which progresses into a major pharyngitis with fever, exudate, leukocytosis and tender regional lymph nodes. An autoimmune reaction develops against joint tissue. A scarlet fever rash may develop in susceptible individuals if the bacteria produces an erythrogenic toxin. Penicillin G is the drug of choice in the treatment of Streptococcus pyogenes infections.

Other toxins produced by Streptococcus pyogenes includes streptokinase, a plasminogen activator; hyaluronidase, which dissolves connective tissue; streptolysin O, which lysis red blood cells and is used as a diagnostic parameter for streptococcal infections; streptodornase, a DNA nuclease; and streptolysin S, another β-hemolysin.

The Streptococci are grouped based of the carbohydrate characteristics of the surface antigens called the Lancefield groupings. Group A are β-hemolytic and, by definition, sensitive to bacitracin. Other members of the genus are resistant to bacitracin and vary in bile solubility and ability to grow in high NaCl media. M protein found on the surface of the bacteria has antiphagocytic activity and therefore serves as a virulence factor. The Group A streptococci are positive for the M protein.

Items 103-114

Many bacteria are associated with commonly occurring illnesses. Routine microscopic analysis of cultures, in many cases, forms the basis for highly reliable preliminary diagnosis of infections. Match the following brief descriptions of bacteria structure and illnesses with the most commonly associated **gram-positive bacteria**.

(A) Actinomyces israelii
(B) Bacillus anthracis
(C) Bacillus cereus
(D) Bacillus subtilis
(E) Bacillus thuringiensis
(F) Clostridium botulinum
(G) Clostridium difficile
(H) Clostridium perfringens
(I) Clostridium tetani
(J) Corynebacterium diphtheriae
(K) Lactobacillus acidophilus
(L) Listeria monocytogenes
(M) Staphylococcus aureus
(N) Staphylococcus epidermidis
(O) Staphylococcus enterocolitis
(P) Streptococcus pyogenes Group A
(Q) Streptococcus pneumoniae
(R) Streptococcus mutans
(S) Mycobacterium avium-intracellulare
(T) Mycobacterium leprae
(U) Mycobacterium tuberculosis
(V) Nocardia asteroides

103. Four members of the same family consumed home made canned beans and became extremely sick and experienced a slight paralysis. The toxin is produced under anaerobic conditions by these club-shaped organisms.

104. A 10 year-old boy had his semiannual dental cleaning. The dentist found some tooth decay. Which bacteria are frequently associated with dental caries?

105. An 82 year-old man was admitted to a U.S. hospital in the Philippines. He showed numerous cuts, burns and bruises on his hands and face. He appeared insensitive to pain in these areas. Tissue biopsy showed acid-fact bacteria.

106. A group of pre-schoolers attending a day care facility all developed fevers and severe sour throats. Cultures of throat swabs grew up β-hemolytic cocci which were bacitracin-snesitive.

107. A middle aged woman cut her scalp while working in her attic. The cut was improperly treated and became purulent. White colonies, of clumped cocci, causing no hemolysis, were recovered on blood agar. The organisms failed to clot plasma.

108. Two workers on a small ranch discovered sick and dying cattle in the field. One developed vesicular lesions on his hands which formed into black eschars. A stained specimen showed large rods in chains with some bearing spores.

109. A camper in a remote region was involved in a shooting accident. He dressed his leg wound and attempted a 5-day hike for assistance. The wound became inflamed, developed a foul odor and discoloration. Later, physicians detected club-shaped bacteria.

110. Bacterial spores were isolated from an over-the-counter organic pesticide used to control Japanese beetles. It was shown not harmful to pets, adults or children.

111. An 8 year-old girl was diagnosed with meningitis. A Sample of her CSF was cultured on blood agar and produced minute, catalase-positive, clear colonies with β-hemolysis. Further tests showed motile rod-shaped organisms.

112. Researchers attempted to reestablish the proper pH and normal vaginal flora by introducing this organism. The rod-shaped bacteria produce lactic acid and are found in healthy women.

113. A scalp wound on a 45 year-old indigent woman became infected. A culture of the pus revealed clustered cocci which produced complete hemolysis on blood agar plates. A coagulase test was positive.

114. The X-ray of a 54 year-old indigent man showed the classic Ghon complex of fibrous lesions. Sputum cultures produced yellowish colonies by the Ziehl-Neelsen procedure.

ANSWERS AND TUTORIAL ON ITEMS 103-114

The answers are: **103-F; 104-R; 105-T; 106-P; 107-N; 108-B; 109-H; 110-E; 111-L; 112-K; 113-M; 114-U**. Bacillus anthracis (B) is often associated with bovine and equine disease. Spores germinate in open wounds or lungs after being inhaled. Vegetative rods produce potent exotoxin that causes tissue necrosis.

Bacillus thuringiensis (E) is commonly used as a biological insecticide. The bacteria produce toxins which prevent the morphogenesis of a variety of insect larvae. The bacteria are harmless to humans.

Clostridium botulinum (F), under alkaline anaerobic conditions, produces a potent toxin in improperly prepared foods. Untreated food poisoning can cause paralysis or death. Heat at 100°C for 10 min will inactivate the toxin.

Clostridium perfringens (H) causes gas gangrene. The spores germinate in uncleaned wounds and produce α-toxin which kills tissue cells. The effects are treatable with antitoxin.

Lactobacillus acidophilus (K) is found in the normal vaginal flora in most women. It produces lactic acid from vaginal glycogen. The resultant lowered pH helps to maintain the proper vaginal flora.

Listeria monocytogenes (L) is a motile rod which produces hemolysin detectable on blood agar. It is associated with meningitis and neonatal infections. A positive catalase reaction and morphology distinguishes it from the streptococci.

Staphylococcus aureus (M) is often found in uncleaned wound infections. It produces grape-like clusters on Gram stain and form complete hemolytic, golden-yellow colonies on blood agar. A positive coagulase test distinguishes it from Staphylococcus epidermidis.

Staphylococcus epidermidis (N) produces white colonies on blood agar. It is coagulase negative. Found on many body surfaces, it is morphologically identical to Staphylococcus aureus.

Streptococcus pyogenes Group A (P) causes strep throat, scarlet fever, rheumatic fever, impetigo and other diseases. Bacitracin sensitivity distinguishes it from other streptococci. The presence of antistreptolysin O in serum is diagnostic.

Streptococcus mutans (R) is a primary cause of dental caries. Colonization is dependent on adherence to enamel by polyglucan synthesis.

Mycobacterium leprae (T) causes leprosy. Advanced stages of leprosy result in the loss of nerve function leading to self-mutilation. The acid-fast bacilli grow slowly on Ziehl-Neelsen medium.

Mycobacterium tuberculosis (U) is the agent of tuberculosis. Primary infection in lungs is confirmed by X-ray. Granulomas occur in both lungs with walled-in bacteria. Reactivation may occur due to malnutrition, alcoholism, diabetes or other life-changing events.

Items 115-125

Many bacteria are associated with commonly occurring illnesses. Routine microscopic analysis of cultures, in many cases, forms the basis for highly reliable preliminary diagnosis of infections. Match the following brief descriptions of bacterial structure and illnesses with the most commonly associated **gram-negative and gram-variable bacteria**.

(A) Bacteroides fragilis
(B) Bordetella pertussis
(C) Borrelia burgdorferi
(D) Borrelia recurrentis
(E) Brucella abortus
(F) Campylobacter fetus
(G) Campylobacter jejuni
(H) Chlamydia psittaci
(I) Chlamydia trachomatis
(J) Coxiella burnetii
(K) Escherichia coli
(L) Klebsiella pneumoniae
(M) Gardnerella vaginalis
(N) Haemophilus ducreyi
(O) Haemophilus influenzae
(P) Legionella pneumophila
(Q) Leptospira interrogans
(R) Mycoplasma hominis
(S) Mycoplasma pneumoniae
(T) Neisseria gonorrhoeae
(U) Neisseria meningitidis

115. A 42 year-old woman with cervicitis was negative when a cervical smear was cultured and Gram stained for Neisseria. A similar sample plated on mouse fibroblast cells and Giemsa stained, showed cytoplasmic inclusions containing minute cocci.

Chlamydia

116. A Chinese-American undergraduate student noticed a large red blotch on her leg. Two weeks earlier she removed tiny deer ticks from her body following a camping trip. Tissue biopsy showed spirochetes.

B. burgdorferi

117. The raw wool processor developed high fever and loss of concentration. Weil-Felix reaction on serum was negative. Serological tests for the agent of Q fever was positive.

C. burnetii

118. A new mother developed a high fever following the birth of her first child. Cord blood cultures where negative on blood agar but positive on Eaton's agar.

119. The Gram stain of urethral discharge smear showed the presence of pink diplococci inside and surrounding polymorphonuclear leukocytes. Oxidase-positive white colonies grew on Thayer-Martin agar.

120. A 10 year-old boy was diagnosed with meningitis. His CSF showed no growth on blood agar. White colonies grew on chocolate agar. Both X- and V-factors were required for colony growth. The Gram stain showed rods.

121. A 19 year-old female college student described a vaginal infection to her doctor. White colonies were recovered on blood and chocolate agar. A Gram stain showed variable coccobacilli.

122. Several children in a first grade class developed diarrhea that lasted 2-5 days. One type of organism grew best under microaerophilic conditions. It was oxidase-positive and showed curved rods in the Gram stain.

123. A bronchial lavage specimen from a 56 year-old man with pneumonia produced creamy white colonies on blood agar. On Endo agar, mucoid, stringy colonies were found. Gram stain showed rods with halos surrounding them.

124. An abscess developed following abdominal surgery on an elderly woman. Nothing grew on blood, nutrient or EMB agar. Abscess fluid plated on blood, nutrient and EMB agar grew no colonies. The specimen showed the presence of coccobacilli. Identical media did produce colonies when incubated in a candle jar.

125. A retired army sergeant contracted pneumonia a week after a visit to Pennsylvania. Microabscesses and inflammatory exudates developed as the disease progressed. Bronchial lavages showed a few pleomorphic thin rods, but cultivation on most media failed to produce any colonies.

ANSWERS AND TUTORIAL ON ITEMS 115-125

The answers are: **115-I; 116-C; 117-J; 118-R; 119-T; 120-O; 121-M; 122-G; 123-L; 124-A; 125-P**. Bacteroides fragilis (A) is the most commonly encountered anaerobic rod-shaped bacterium. It is frequently recovered from abscesses of the intestinal and genital tracts. Some strains will tolerate low oxygen levels.

Borrelia burgdorferi (C) is the spirochete which causes Lyme disease. It in microaerophilic and visualized with the Giemsa and silver stains. Human infection may involve the joints and heart.

Campylobacter jejuni (G) is a major cause of diarrhea associated with the consumption of poorly processed poultry. It produces an enterotoxin. The bacteria are shaped like curved rods.

Chlamydia trachomatis (I) causes sexually transmitted disease along with pneumonia (spread by inhalation) and conjunctivitis (spread by direct contact). It is an obligate intracellular bacteria which requires a eucaryotic host cell in which to grow.

Coxiella burnetii (J) causes Q fever. This rickettsial organism does not require an arthropod vector for transmission. Transmission is by inhalation of contaminated wool or dried fomites.

Klebsiella pneumoniae (L) produces a large amount of capsule polysaccharide. Many serotypes (>70) exist. The capsular properties distinguish it from other bacteria.

Gardnerella vaginalis (M) often causes secondary vaginitis. It is a slow-growing bacteria with growth properties similar to those of Haemophilus sp.

Haemophilus influenzae (O) is a slow-growing bacillus which requires both X (hematin) and V (NAD) factor. Staphylococcus aureus excretes V-factor and will therefore support satellite colonies of H. influenzae on blood agar.

Legionella pneumophila (P) is the cause of Legionnaires' disease. It is a facultative intracellular pathogen of macrophages and monocytes. Colonies grow on charcoal yeast extract agar.

Mycoplasma hominis (R) is the cause of postpartum fever experienced by new mothers. The organism is an opportunist which may infect during child birth. The absence of a cell wall renders the Gram stain uninformative. Speciation is by serotype differences.

Neisseria gonorrhoeae (T) causes gonorrhea. Kidney bean-shaped gram-negative cocci within white cells is diagnostic. Sugar fermentation is used to distinguish different Neisseria species. Thayer-Martin agar restricts growth of other bacteria by the inclusion of various selective antibiotics.

Items 126-136

Many bacteria are associated with commonly occurring illnesses. Routine microscopic analysis of cultures, in many cases, forms the basis for highly reliable preliminary diagnosis of infections. Match the following brief descriptions of bacteria structure and illnesses with the most commonly associated **gram-negative and gram-variable bacteria**.

(A) Pasteurella multocida
(B) Proteus mirabilis
(C) Proteus vulgaris
(D) Pseudomonas aeruginosa
(E) Pseudomonas pseudomallei
(F) Rickettsia prowazekii
(G) Rickettsia rickettsii
(H) Rickettsia tsutsugamushi
(I) Salmonella enteritidis
(J) Salmonella typhi
(K) Shigella dysenteriae
(L) Shigella sonnei
(M) Treponema pallidum
(N) Treponema pertenue
(O) Ureaplasma urealyticum
(P) Vibrio cholerae
(Q) Vibrio parahemolyticus
(R) Yersinia enterocolitica
(S) Yersinia pestis

126. A 32 year-old man experienced diarrhea, cramps and bloody squirty stools following a business encounter group session in a campsite area. Stool specimen plated on MacConkey agar produced white colonies which gram-stained as negative rods.

127. A teenage boy developed a urethritis and went to the clinic. The Gram stain, chocolate agar and blood agar were all negative. Specimen plated on Eaton's agar (supplemented with urea) produced minute colonies after 2 weeks incubation.

128. A male hunter noticed a rash on his legs that spread to his hands. The first rash appeared where he had removed ticks after an outing in the woods. A serum sample was positive for all antigens in the Weil-Felix test.

129. A 12 year-old boy stepped on broken glass and received medical attention. However he did not complete the follow-up directions and the wound became infected. A lactose non-fermenting rod was isolated. It was resistant to many antibiotics.

P. aeruginosa

130. A woman received a deep claw gash from her pet cat. The wound became infected within 24 hrs. A fast-growing, oxidase-positive rod was isolated.

Pasteurella multocida

131. A young boy near Denver, Colorado, was playing with a dead prairie dog. Several days later he became sick and his condition was diagnosed as bubonic plague.

Y. pestis

132. The serum of a traveling salesman was positive in the VDRL test and the FTA-ABS test.

T pallidum

133. Construction workers became sick with diarrhea, stomach cramps and vomiting. The isolate was lactose nonfermenting, motile and produced hydrogen sulfide.

Salmonella enteritidis

134. Bacteria from a urinary tract infection produced swarming colonies of blood agar but not MacConkey agar. The isolate produced hydrogen sulfide, was urease-positive but was indole-negative.

P. mirabilis

135. A refugee from India developed severe diarrhea which produced rice water stools with no accompanying blood. A comma-shaped bacteria was isolated which grew faster under alkaline conditions.

Vibrio cholerae

136. The retired army soldier had fever and headaches, symptoms which were similar to the epidemic typhus he experienced years earlier. He had been confined to a hospital nursing home for the previous four years.

R. prowazekii = typhus + Brill's disease

ANSWERS AND TUTORIAL ON ITEMS 126-136

The answers are: **126-L; 127-O; 128-G; 129-D; 130-A; 131-S; 132-M; 133-I; 134-B; 135-P; 136-F**. Pasteurella multocida (A) is associated with dog or cat bites or scratches. The fast growing organism produces toxic effects within 24 hrs of the encounter. The bacteria ferments many sugars and gives an oxidase-positive reaction.

Proteus mirabilis (B) infects the urinary tract. It has a characteristic swarming growth pattern that is inhibited by the bile in MacConkey agar. It is urease- and hydrogen sulfide-positive, but indole-negative (which distinguished it from Proteus vulgaris which is indole-positive).

Pseudomonas aeruginosa (D) is a lactose nonfermenting rod. It is an opportunist which is resistant to many antibiotics and capable of growing in superficial and deep wounds. It is also responsible for major nosocomial infections.

Rickettsia prowazekii (F), the agent of epidemic typhus, also causes Brill's disease which is a reemergence of a milder form of the disease, years later, in the absence of new vector (louse) contact.

Rickettsia rickettsii (G) is the cause of Rocky Mountain spotted fever. The characteristic rash is more pronounced on the extremities than on the trunk. Antibody reacts to all of the Proteus antigens of the Weil-Felix test. Transmission occurs via tick bite.

Salmonella enteritidis (I) causes profound diarrhea and vomiting within hours of consuming contaminated foods. Groupings are based on surface O-antigens and H-antigens located on the flagella. Hydrogen sulfide production distinguishes it from the Shigella species.

Shigella sonnei (L) can cause diarrhea in which the watery stools change from clear to bloody as the disease progresses. S. sonnei (Group D) is more frequently encountered than S. dysenteriae (Group A).

Treponema pallidum (M) causes syphilis. The VDRL test is a flocculation test which indirectly detects anti-treponemal antibody using nonspecific antigen. The fluorescent treponemal antibody absorption test (FTA-ABS) tests for antibody directed against the bacterial antigen.

Ureaplasma urealyticum (O) is a mycoplasmal organism which causes venereal infections. It is the slowest growing of the mycoplasmas and requires urea and cholesterol for growth. Eaton's agar is an enriched nutrient system with inhibitors of faster growing bacteria.

Vibrio cholerae (P) causes classical cholera. The comma-shaped rods grow on many nutrients agars but faster under alkaline conditions with citrate, bile and thiosulfate

Yersinia pestis (S) is the agent of bubonic plague. The disease is transmitted in the prairie dog population, carried by fleas. Accidental infection to humans occurs through contact with infected animals. The infected swollen lymph node (bubos) commonly occurs in the groin area.

Items 137-143

Antibiotics exert either bacteriostatic or bacteriocidal actions on bacteria through different mechanisms. Match the inhibitory mechanisms in the items below with the most appropriate antibiotics in the answers.

(A) Penicillins
(B) Tetracycline
(C) Chloramphenicol
(D) Streptomycin
(E) Gentamycin
(F) Erythromycin
(G) Sulfonamides
(H) Cephalosporins
(I) Fluoroquinolones

137. Antibiotics consisting of a 5-member thiazolidine ring. They inhibit gram-positive bacteria by preventing cell wall synthesis by interfering with the transpeptidation reaction.

138. Antibiotics consisting of a 6-member dihydrothiazine ring. They inhibit both gram-positive and gram-negative bacteria by preventing cell wall synthesis in growing bacteria.

139. Bactericidal, aminoglycoside, which binds to the 30S ribosomal subunits and prevent protein synthesis.

140. Broad spectrum antibiotic which bind to the 30S ribosomal subunit and block attachment of aminoacyl-tRNA.

141. Broad spectrum antibiotic which binds to the 50S ribosomal subunit by inhibiting peptidyl transferase.

142. A macrolide that blocks protein synthesis by binding to the 50S ribosomal subunit blocking the translocation step.

143. Analogues of para-aminobenzoic acid (PABA) that block folic acid synthesis.

ANSWERS AND TUTORIAL ON ITEMS 137-143

The answers are: **137-A; 138-H; 139-I; 140-B; 141-C; 142-F; 143-G**. Penicillins (A) are the prototype β-lactam antibiotics. They are derived from the mold Penicillium. They interfere with the transpeptidation that results in the cross-linking of the glycan chains in the bacterial cell wall. The addition of the cross-bridge is blocked. A weakened bacterial cell wall results which ruptures under osmotic imbalance.

Tetracycline (B) is effective against many gram-positive and gram-negative bacteria. It binds to the 30S ribosomal subunit of bacteria and blocks tRNA interaction with mRNA in a reversible reaction.

Chloramphenicol (C) is produced by Streptomyces. It binds to the 50S ribosomal subunit of bacteria and prevents peptide synthesis. The effects are reversible.

Streptomycin (D) inhibits bacterial protein synthesis by binding to the bacterial 30S ribosomal subunit. The compound is poorly absorbed by many bacteria and therefore is less effective.

Gentamycin (E), like streptomycin, binds to the 30S ribosomal subunit of bacteria. It has a wide spectrum of activity and is heat stabile.

Erythromycin (F) is a macrolide that binds to the 50S ribosomal subunit of bacteria. It also blocks the translocation of peptidyl-tRNA.

Sulfonamides (G) are analogues of para-aminobenzoic acid (PABA). They compete for the enzyme, dihydropteroate synthetase, in the synthesis of folates and folic acid. The latter is an essential coenzyme in protein and nucleic acid synthesis.

Cephalosporins (H) are produced by the fungus Cephalosporium. They act similarly to the penicillins but differ in structure. Cephalosporins inhibit bacteria by interfering with bacterial cell wall synthesis.

Fluoroquinolones (I) resemble nalidixic acid but with the addition of a piperazine substitute. These compounds inhibit both gram-positive and gram-negative organisms by binding to DNA gyrase. They therefore inhibit DNA replication.

Items 144-150

A group of 25 fifth-grade students had lunch at a fast food restaurant. Most ate hamburgers and french fries prepared quickly for the noontime rush. The next day most of the children experienced severe bloody diarrhea and nausea with several requiring hospitalization. Stool and blood cultures recovered small, gram-negative, nonmucoid bacilli which fermented lactose, were indole-positive, produced acid and carbon dioxide but produced neither hydrogen sulfide nor urease.

144. The organism most likely responsible for the intestinal infection is

(A) Pseudomonas aeruginosa
(B) Proteus vulgaris
(C) Klebsiella pneumoniae
(D) Vibrio cholerae
(E) Escherichia coli

145. Serotyping of these organisms is based primarily on the

(A) capsule antigens
(B) flagella (H) antigens
(C) somatic (O) and (H) antigens
(D) cell membrane antigens
(E) cytoplasmic antigens

146. The K (pili) antigens may bind to a host cell ligand ending in

(A) glucose
(B) fructose
(C) maltose
(D) mannose
(E) galactosamine

147. The organism isolated from the children might be best categorized as

(A) enterotoxigenic
(B) enteroinvasive
(C) enterohemorrhagic
(D) enteropathogenic
(E) enterogenic

148. The potent toxin expressed by this diarrhea-producing isolate is due to the expression of

(A) LT (heat labile toxin)
(B) ST (heat stabile toxin)
(C) Shigella dysenteriae-like toxin type 1
(D) Salmonella-like enterotoxin
(E) Vibrio-like toxin

149. The primary inhibitory action of the toxin is

(A) protein synthesis inhibition
(B) adenylate cyclase inhibition
(C) guanylate cyclase stimulation
(D) DNA synthesis inhibition
(E) RNA synthesis inhibition

150. Rapid clinical identification of the extremely virulent strains of the organism have recently relied on

(A) DNA fingerprinting
(B) agglutination tests
(C) DNA probes
(D) rabbit ileal loop technique
(E) sugar fermentation assay

ANSWERS AND TUTORIAL ON ITEMS 144-150

The answers are: **144-E; 145-E; 1460-D; 147-C; 148-C; 149-A; 150-C**. Intestinal infections caused by pathogenic strains of Escherichia coli have remained a major public health concern due to the like-threatening nature of some infections. E. coli are easily identified by their ability to grow on a number of synthetic media. Lactose fermentation with the production of acid and gas are characteristic identifiers. These organisms produce neither urease nor hydrogen sulfide which distinguishes them from many Proteus species.

Typing of E. coli isolates is dependent on the somatic O antigens and flagellar H antigens. The pili (K) antigen may act as an adhesin which enables the organism to adhere and colonize the intestinal tract. A terminal mannose sugar residue on the mucosal surface is an important ligand in the attachment of bacteria to host cells.

The virulence of the many different strains is dependent on the type of plasmid or virus-controlled toxin. The heat-labile (LT) and heat-stabile (ST) toxins are both involved in diarrhea. They are found in the enterotoxigenic E. coli (ETEC) strains. The enteroinvasive E. coli (EIEC) strains rarely cause diarrhea. The enteropathogenic E. coli (EPEC) strains produce diarrhea and both ST and LT.

The most virulent of these organisms belong to the enterohemorrhagic E. coli (EHEC) strains. These express LT, ST and a very potent Shigella dysenteriae-like type 1 toxin. Bloody diarrhea is a characteristic of this group of bacteria. Advanced infection may prove fatal. The O157:H7 serotype is the most virulent EHEC since it expresses all three toxins. The Shigella dysenteriae-like type 1 toxin inhibits host cell protein synthesis resulting in cell death.

Rapid typing of E. coli isolates has been advanced through the use of DNA probes which identify those strains expressing the O157:H7 antigens. Aggressive treatment with antibiotics is important with individuals infected with EHEC strains.

Items 151-155

Patients who attended a small dental clinic were concerned when it was reported that a patient had been infected with the AIDS virus from contaminated dental instruments. The dentist assured the patients that he practiced correct procedures for cleaning and disinfecting his equipment. Sterilization of equipment was not performed between each patient.

151. Effective sterilization requires which of the following conditions?

(A) 70°C for 30 minutes
(B) 100°C for 10 minutes
(C) 70% alcohol solution
(D) 121°C, 15 psi, for 20 minutes
(E) 100°C for 30 minutes

152. Which of the following gases have been used to sterilize heat-sensitive materials?

(A) carbon monoxide
(B) carbon dioxide
(C) nitric oxide
(D) ethylene oxide
(E) nitrous oxide

153. Ultraviolet light inactivates vegetative bacteria by

(A) inducing single-stranded breaks in DNA
(B) by inducing double-stranded breaks in DNA
(C) inducing thymine dimers
(D) denaturing nucleoproteins
(E) cross-linking sugar groups

154. Ionizing radiation inactivates vegetative bacteria by

(A) inducing single-stranded breaks in DNA
(B) inducing double-stranded breaks in DNA
(C) inducing thymine dimers
(D) denaturing nucleoproteins
(E) cross-linking sugar groups

155. An effective membrane pore size necessary to filter out most rigid bacteria is approximately

(A) 1.00 μm
(B) 0.5 μm
(C) 0.2 μm
(D) 0.02 μm
(E) 0.001 μm

ANSWERS AND TUTORIAL ON ITEMS 151-155

The answers are: **151-D; 152-D; 153-C; 154-A; 155-C**. There are fundamental differences between the cleaning, disinfecting and sterilizing of dental equipment. Cleaning implies the removal of significant amounts of undesired materials from a surface; disinfection involves the destruction of most microorganisms by chemical means; and sterilization implies the destruction of all microorganisms. A disinfectant may not destroy all forms of a pathogen. The effectiveness of chemical disinfectants varies with the active agents used in the formulation. New chemicals are introduced annually which promise extended bactericidal, sporicidal and viricidal activities.

Sterilization is best accomplished through the application of moist heat. The autoclave presents conditions under which all known forms of infectious agents including vegetative bacteria, viruses, fungi and spores are inactivated. A temperature of 121°C and 15 pounds per square inch (PSI) for 15 minutes is sufficient to kill all known pathogens. The moisture of the autoclave system ensures penetration of spores, and viruses. Most enzymes and nucleic acids are also denatured under these conditions. Dry heat systems require higher temperatures (160°C) and much longer exposures. Not all materials or substances can withstand autoclave conditions and remain functional, therefore alternative sterilization methods are used. Alcohols (70-90%) which denature proteins are ineffective against spores and many viruses.

Heat-sensitive materials may be sterilized by gas. Ethylene oxide (10%) in freon or carbon dioxide has been used to sterilize plastics. Effective sterilization requires several hours of exposure to this toxic gas.

Irradiation such as ultraviolet light and ionizing irradiation came effectively sterilize materials. Ultraviolet (UV) light induces thymine dimers in nucleic acids. Systems which lack adequate DNA repair systems are inactivated. Ionizing irradiation such as gamma rays cause single-stranded nucleic acid breaks. Biological systems which lack adequate repair enzymes may incorporate incorrect nucleotides resulting in lethal mutations or incomplete protein products formed. The effective penetrating power of UV is less than ionizing irradiation.

Solutions are often sterilized by filtration. Most rigid bacteria are effectively removed if a solution is passed through a sterilized membrane which has a pore size of 0.2 μm. Viruses and mycoplasmas easily pass through this barrier and therefore might require other sterilization procedures.

Items 156-161

A 75 year-old man lived in a retirement village. He received periodic medical check-ups and was in relatively good health. During the winter an outbreak of influenza affected many residents of the complex even though all had received vaccinations. The man developed a progressive lower respiratory tract illness with symptoms including abrupt chills, high fever, productive coughs and chest pains. Sputum cultures revealed a large number of gram-positive diplococci which were also found in blood samples. Chest X-rays showed lobar consolidation. The man's conditioned deteriorated ending with his death.

156. The organism observed in the patient's sputum and blood probably belongs to which of the following genera:

(A) Staphylococcus
(B) Streptococcus
(C) Micrococcus
(D) Neisseria
(E) Candida

157. The organism streaked on a blood agar plate would produce

(A) β-hemolytic, bacitracin-sensitive, colonies
(B) γ-hemolytic, optochin-resistant, colonies
(C) α-hemolytic, optochin-sensitive, colonies
(D) β-hemolytic, bacitracin-resistant, colonies
(E) α-hemolytic, optochin-resistant, colonies

158. Serotyping of the bacteria is based on

(A) phage typing
(B) precipitation by antisera
(C) flagella labeling by antisera
(D) antisera swelling of the capsule
(E) DNA fingerprint analysis

159. Vaccines against the organism are prepared against the

(A) cell wall
(B) flagella
(C) capsule
(D) ribosomes
(E) nucleic acid proteins

160. A drug of choice often prescribed in treating infections caused by this organism is

(A) streptomycin
(B) sulfonamide
(C) penicillin
(D) tetracycline
(E) cephalosporin

161. A toxin isolated from animals infected experimentally with the organism is

(A) streptolysin O
(B) pneumolysin
(C) erythrogenic toxin
(D) streptolysin S
(E) hyaluronidase

ANSWERS AND TUTORIAL ON ITEMS 156-161

The answers are: **156-B; 157-C; 158-D; 159-C; 160-C; 161-B**. The bacteria isolated from sputum and blood belong to the genus Streptococcus. Some members of the genus such as Streptococcus pneumoniae, also called pneumococci, form short chains in which the bacteria appear paired. Each gram-positive coccus is slightly elliptical. Overgrown cultures will stain variably as the bacteria loose viability. S. pneumoniae plated on blood agar forms clear flat colonies surrounded by zones of α-hemolysis. The organism is sensitive to optochin which distinguishes it from the viridans streptococci which also produce α-hemolysis on blood agar.

The quellung (capsule swelling) reaction has been used to serotype the pneumococci. The test uses antisera prepared against known serotypes of S. pneumoniae. A positive antibody-antigen reaction results in the swelling of the polysaccharide capsule which surrounds the organisms. There are over 80 capsule types. The antigenic properties of the capsule polysaccharide is the basis of pneumococcal vaccines.

Pneumococcal infections in either elderly or immunosuppressed patients are a major health concern since they are often life-threatening. The traditional drug of choice has been penicillin, however new drug-resistant strains have emerged. Chloramphenicol and erythromycin have been used as secondary drug choices. Virulence factors expressed in the pneumococci include pneumolysin which has enzymatic properties similar to neuraminidase. It cleaves sialic acid residues from glycoproteins and glycolipids, however its role in pathogenesis is still unclear. The ability of the polysaccharide capsule to prevent phagocytosis is an important virulence factor.

Pneumococcal disease in the elderly patient causes respiratory tract failure by rapid growth in the alveoli, causing fluid production and white cell infiltration. Bacteremia may also cause endocarditis, meningitis, arthritis or peritonitis.

Items 162-164

A young mother became concerned when her 24 month-old infant daughter developed a mild fever and cried continuously. The infant drooled excessively with difficulty sleeping and eating. The pediatrician, bowing to the mothers's requests, performed bacterial cultures of the infant's blood, saliva, urine and stool specimen. The symptoms continued for several days before slowly subsiding.

The following organisms were isolated from the infants specimen:

Blood:	NEGATIVE
Saliva:	viridans streptococci, Lactobacillus sp.
Urine:	NEGATIVE
Stool:	Escherichia coli, Lactobacillus sp., Bacteroides sp., Clostridium sp.

162. The bacteriological findings are consistent with

(A) possible enteric infection
(B) hemolytic streptococcal infection
(C) bacteremia
(D) respiratory tract infection
(E) normal flora

163. All of the following are considered normal microbial flora of the colon **EXCEPT**:

(A) Bacteroides sp.
(B) Fusobacterium sp.
(C) Clostridium sp.
(D) Escherichia sp.
(E) Vibrio sp.

164. All of the following specimens are normally free of bacteria in most healthy individuals **EXCEPT**:

(A) blood
(B) urine
(C) cerebrospinal fluid
(D) muscle tissue
(E) tears

ANSWERS AND TUTORIAL ON ITEMS 162-164

The answers are: **162-E; 163-E; 164-E**. Occasionally the disposition or transient irritability of an individual, especially of infants, might be interpreted as symptoms of microbial infection. The bacteriological findings in specimens taken from the infant are consistent with normal flora. Many body sites are colonized by avirulent organisms or opportunistic bacteria. These organisms are blocked from entering the body by epithelial barriers such as the epidermis. Staphylococcus aureus isolated from the surface of the skin is not an unusual finding; however, its presence in wounds, blood or inner ear infections is cause for alarm and warrants antibiotic treatment.

All of the bacteria isolated from the stool specimen are normal, however organisms such as Vibrio sp. could represent possible enteric infection. Bacteria account for approximately 25% of the mass of human adult feces. Infants, since they consume human or animal milk, often harbor large numbers of Lactobacillus sp. in their saliva and stools.

Body specimens such as blood, urine, CSF and tissues are normally free of bacteria. Tears, however contain low numbers of bacteria even in the presence of lysozyme. Constant exposure to the environment accounts for the ever present low number of bacteria in tears.

The infant was free of bacterial infection based on the specimens tested. Factors such as her general disposition, teething or family tensions may have accounted for the observed symptoms. Similarly, a nonbacterial factor could have been responsible. This seems unlikely because the symptoms abated soon after examination and required only supportive treatment from the pediatrician.

CHAPTER II

VIROLOGY

Items 165-168

A 22 year-old homosexual male had severe colds accompanied by copious bloody sputum with greenish flecks, fever, overall loss of energy and skin rashes. He underwent a physical examination which included a full blood workup. His overall cell count was within the normal range with the exception of his CD4+ cells which was 150/ mm^3 (normal range 300 - 1200).

165. The heavy sputum may have been due to an opportunist infection caused by

(A) Chlamydia pneumoniae
(B) Streptococcus pneumoniae
(C) Mycoplasma pneumoniae
(D) Mycobacterium tuberculosis
(E) influenza A virus

166. The patient's profile best fits that of an individual infected with

(A) HIV
(B) Epstein-Barr virus
(C) influenza A virus
(D) coronavirus
(E) adenovirus

167. Surrogate markers used in the diagnosis of the above illness include all of the following **EXCEPT**:

(A) CD4+ cell number
(B) β_2-microglobulin
(C) neopterin
(D) delayed-type hypersensitivity
(E) fever

168. Drugs used in the treatment of the pathogen include all of the following **EXCEPT**:

(A) AZT
(B) ddI
(C) ddC
(D) IFn-α
(E) acyclovir

ANSWERS AND TUTORIAL FOR ITEMS 165-168

The answers are: **165-D; 166-A; 167-E; 168-E**. Infection by HIV and the disease AIDS are not easily diagnosed since the incubation period may vary from months to years. Most diagnostic procedures for the virus detect circulating antibodies which emerge slowly in some cases. Opportunistic infections, such as Mycobacterium tuberculosis occur at a higher rate in immunocompromised individuals. Deep productive coughing with blood are indications of possible mycobacterial infection and therefore warrants further confirmatory tests.

Surrogate markers for AIDS take into account several factors since the CD4+ (helper) T cell count fluctuates among individuals. A count of less than 300 cells/mm^3 is considered a sign of immunosuppression. β_2-microglobulin, a component of the major histocompatibility complex; neopterin, produced by white cells; and interferon are all elevated in HIV-infected people. Delayed-type hypersensitivity (DTH) is suppressed. Fever may be a result of pathology due to opportunistic infections and therefore is not a reliable surrogate marker for HIV infection.

AZT, ddI, ddC and interferon have all been used in different treatment protocols for AIDS. Acyclovir is effective against certain DNA viruses such as herpes simplex, but not effective against HIV.

Item 169-172

An enzyme-linked immunosorbent assay (ELISA) for the hepatitis B virus (HBV) was performed on serum from 12 different undergraduates. Only students A and B were positive for IgG against the surface antigen of the virus. Student A was from Nigeria and student B was a United States citizen. Both were similar in age, race and overall good health, but from different cultural lifestyles.

169. The initial ELISA test for exposure to the hepatitis B virus is limited in that it

(A) cross reacts with many other viral infections
(B) does not distinguish HBV from other hepatitis viruses
(C) does not distinguish between HBV-infected individuals and those vaccinated against HBV
(D) requires a large sample (>20 cc) of serum
(E) is less sensitive than a complement fixation assay

170. A test that would distinguish a recent exposure to HBV from long-past (9 months) exposure to the virus could involve detection of

(A) anti-core (IgG) antibody
(B) anti-polymerase (IgG) antibody
(C) anti-core (IgM) antibody
(D) anti-e (IgG) antibody
(E) anti-x (IgG) antibody

171. **IF** student A had been vaccinated for HBV, and had not been exposed to the active virus, his antibody titer nine (9) months later would most likely show only antibody against the

(A) core antigen
(B) polymerase antigen
(C) envelope lipid
(D) nucleic acid
(E) surface antigen

172. Known mechanisms of transmission of the HBV includes all of the following **EXCEPT**:

(A) contaminated blood
(B) organ transplants
(C) penile-vaginal sex
(D) penile-anal sex
(E) aerosol (droplet) inhalation

ANSWERS AND TUTORIAL FOR ITEMS 169-172

The answers are: **169-C; 170-C; 171-E; 172-E**. Individuals infected with the hepatitis B virus may undergo an acute infection which produces the release of large numbers of infectious virions (Dane particles) into the blood. Antibody (IgG) against the surface antigen of the virus has neutralizing properties and appears 4-6 months after exposure. It remains circulating in the blood for over a year. Routine ELISA screening tests for the presence of anti-surface antibody will not distinguish the seroconverting antibody elicited by HBV vaccines, since the latter use recombinant surface antigen. Acquired infections will produce, in addition to antibody to the surface antigen, in most cases, antibody to other viral structural components such as the core, polymerase and e antigens. Such responses would not occur in the vaccinated individual. The expression of IgM antibody against viral structural components is a short-lived immune response and therefore may be used to distinguish between recent versus "old" infections.

HBV is endemic in many areas of the world. Many countries of the sub-Saharan region of Africa show infection rates in excess of 80% of the population. Most are infected early in life and express the complicating incidence of higher frequencies of hepatic cell carcinomas. Infections occur through contaminated blood products and sexual intercourse. The virus is not excreted through the respiratory tract.

<u>Item 173-178</u>

A 4 year-old African-American male developed a fever, sore throat and unusual irritability. A physical examination revealed 15-25 small, gray-white, shallow, ulcerative lesions along the posterior margin of his soft palate. The pharyngeal walls were extremely red. There were no lesions on the gum but a few on the anterior part of the oral cavity. The child was dark-complexioned and had no obvious rash.

173. The most likely diagnosis is

(A) herpetic gingivostomatitis
(B) measles
(C) chickenpox
(D) herpangina
(E) hand, foot and mouth disease

174. The causative viral agent is most likely a

(A) herpes simplex virus
(B) enterovirus
(C) paramyxovirus
(D) varicella-zoster virus
(E) rhabdovirus

175. The lesions found in the child's mouth are called

(A) pustules
(B) Koplik's spots
(C) Negri bodies
(D) Byran's spots
(D) pocks

176. Another name for the disease is

(A) rubella
(B) roseola infantum
(C) rubeola
(D) relapsing fever
(E) rheumatic fever

177. Transmission of the disease occurs primarily through

(A) intimate contact
(B) inhalation of aerosol droplets
(C) ingestion of contaminated foods
(D) the bite of an arthropod vector
(E) blood transfusions or other blood products

178. A possible complication associated with infection by this virus includes all of the following **EXCEPT**:

(A) encephalitis
(B) otitis media
(C) sterility
(D) bacterial superinfections
(E) diarrhea

ANSWERS AND TUTORIAL FOR ITEMS 173-178

The answers are: **173-B; 174-C; 175-B; 176-C; 177-B; 178-E**. Measles virus is a highly infectious childhood virus usually attacking youngsters 6 months and over. The virus belongs to the paramyxovirus group which consists of enveloped RNA viruses which possess strong hemagglutinin activity on their surfaces. Infection usually occurs through the inhalation of contaminated aerosols; however, youngsters are also likely to place contaminated articles (such as toys or shared foods) into their mouths and assist in the transmission of the virus.

Measles undergoes several viremic stages which disseminate the virus throughout the body. The skin rash may vary in intensity in different individuals and even go unnoticed in dark-complexioned individuals. However, characteristic oral lesions, called Koplik's spots develop in the posterior of the mouth and inner cheeks. These represent focal areas of viral infection, cell necrosis and infiltrating white cells. Another name for the measles is Rubeola, which is different from Rubella (caused by a togavirus) and roseola (caused by a herpes virus).

Complications associated with measles include neurological damage (encephalitis), inner ear infections (otitis media), sterility (especially in older individuals), and bacterial superinfections. The latter may be due, in part, to a transient immunosuppression associated with measles infections. The virus does not propagate in the gastrointestinal tract and therefore diarrhea is not a common symptom directly associated with the infection.

Item 179-182

A diagnosis of malaria was made in the case of a 23-year old female who recently immigrated to the United States from Vietnam. During treatment of the parasitic infection she developed shaking chills and spiking fever. Several days later, small clusters of vesicles appeared around her mouth and nostrils.

179. The most likely cause of the facial lesions is

- (A) Coxsackie B virus
- (B) herpes simplex virus
- (C) varicella-zoster virus
- (D) variola virus
- (E) vaccinia virus

180. The most likely site of latency for this virus is the

(A) dorsal root ganglion
(B) sacral ganglion
(C) trigeminal ganglion
(D) adenoid tissue
(E) cornea

181. One method for distinguishing different viruses of the same group could include studying

(A) inclusion formation within infected tissues
(B) neutralizing antibodies
(C) size of the vesicles
(D) location of the vesicles
(E) age of the patient

182. The expression of the virus infection is most likely due to the

(A) normal menstrual cycle of the patient
(B) general health of the patient
(C) antimalarial drug treatments
(D) new lifestyle of the patient
(E) recent acquisition of the virus

ANSWERS AND TUTORIAL FOR ITEMS 179-182

The answers are: **179-B; 180-C; 181-B; 182-C**. Most individuals have been exposed to a herpes simplex virus by their early teen years. The virus, and other members of this group, often enter a latency phase whereby the virus remains dormant within neuronal cells. A variety of circumstances may assist in the activation of the virus and result in the production of lesions, usually within an area of prior lesion expressión. Chemotherapy, sunlight, trauma and even menstrual cycle changes have been reported to have affects of the expression of a latent herpes virus infection. Viruses expressed on the face establish latency within the trigeminal ganglion since its nerve branches innervate the face.

The size, location and patterns of herpes virus lesions are indicative of different infections. Broad variations occur among individuals. Likewise, their ultrastructures are similar and not easily distinguished. A simple distinguishing procedure would be the use of a neutralizing antibody. Recently, the use of nucleic acid probes and DNA fingerprinting techniques have been used to identify herpes viruses.

Items 183-186

There was an outbreak of an illness among approximately 40% of newly recruited enlisted men at Camp Macho in September, 1990. It was characterized by sore throat, conjunctivitis and fever. No common bacteria were isolated from the infected men but a similar virus was isolated from the throat, conjunctiva and stools several individuals. Laboratory tests showed it to be a solvent-resistant DNA virus.

183. **IF** a transmission electron microscopic study of the virions were conducted, which of the following features would you expect to see?

(A) Bullet-shaped virions
(B) Virions with solar corona-like structures on the envelopes
(C) Icosahedral virions with satellite-like attachments
(D) Virions with sand-like particles in the core
(E) Large brick-shaped virions with lateral bodies and a central core

184. The clinical picture, limited viral composition and electron microscopic observations would place it as a member of which of the following groups?

(A) Herpes virus
(B) Poxvirus
(C) Paramyxovirus
(D) Rhabdovirus
(E) Adenovirus

185. Members within the above viral group can be distinguished based on

(A) virulence
(B) electron microscopic ultrastructure
(C) serotypes
(D) animal host range
(E) cytopathic effects

186. Treatment of infections would involve

(A) the administration of antiviral compounds
(B) the treatment of symptoms
(C) vaccination
(D) the administration of immune globulin
(E) irradiation therapy

ANSWERS AND TUTORIAL FOR ITEMS 183-186

The answers are: **183-C; 184-E; 185-C; 186-B**. The adenoviruses are responsible for a large number of different illnesses involving the respiratory tract, gastrointestinal tract, hepatic system and ocular system. Distinguishing feature of these viruses include their resistance to envelope solvents since they lack envelopes, their large, double-stranded DNA genome and their characteristic ultrastructure. Electron micrographs reveal naked icosahedrons with knob-like structures at the tips of each vertex.

Adenoviruses cause conjunctivitis, such as "swimming pool" conjunctivitis, since they are resistant to high concentrations of chlorine used in water treatment. Pharyngitis, fever and acute respiratory illnesses are caused by these viruses. Certain laboratory animals form malignant tumors induced by transforming genes encoded by certain adenoviruses.

There are of 50 adenoviruses which infect humans out of a total of more that 100 different viruses. Strain differences are based of serotypic varieties caused by antigenic properties of the outer capsid.

Most adenovirus infections eventually resolve themselves. There are no reliable treatments for adenovirus infections other than the treatment of the symptoms and prevention of secondary bacterial infections. Vaccines were once available for the general public, but were discontinued. They are now only administered to military populations.

Items 187-192

An Annapolis Naval Academy midshipman developed symptoms consistent with infectious mononucleosis which included fever, lymphadenopathy, sore throat, fatigue and malaise. On the fourth day of the disease he saw a physician who ordered a bacterial throat culture and a heterophile test. Both were negative. The white cell count was 12,000 cells/mm^3 with lymphocytosis and 10% atypical lymphocytes. Liver enzymes were slightly elevated but HB_sAg was negative.

187. Assuming this is a case of infectious mononucleosis, which of the following interpretations regarding the negative heterophile test is **FALSE**?

(A) The etiological agent may be cytomegalovirus
(B) Heterophile antibody has not appeared yet since the infection is new
(C) The sailor may never become heterophile positive because 15-20% of cases of infectious mononucleosis are consistently heterophile antibody negative
(D) The heterophile test is not a highly specific test
(E) The heterophile test measures only IgG

188. Most cases of infectious mononucleosis are caused by the

(A) Epstein-Barr virus
(B) cytomegalovirus
(C) herpes simplex virus-6
(D) adenovirus
(E) Coxsackie virus

189. The circulating cell most infected by the virus is the

(A) helper T cell
(B) cytotoxic T cell
(C) macrophage
(D) B cell
(E) plasma cell

190. Downey cells are atypical

(A) lymphocytes infected by the virus
(B) macrophages infected by the virus
(C) epithelial cells infected by the virus
(D) plasma cells infected by the virus
(E) red blood cells infected by the virus

191. A characteristic pathology of infectious mononucleosis is

(A) transient CMI suppression
(B) transient humoral suppression
(C) significant hemolysis
(D) short term (1-2 weeks) viral shedding
(E) noninfectious blood

192. Transmission of the infection may occur by all of the following mechanisms **EXCEPT**:

(A) blood transfusion
(B) organ transplants
(C) sexual contact
(D) saliva transfer (kissing)
(E) contaminated foods

ANSWERS AND TUTORIAL ON ITEMS 187-192

The answers are: **187-E; 188-A; 189-D; 190-A; 191-A; 192-E**. The Epstein-Barr virus is the causative agent of most cases of infectious mononucleosis. It initially replicates in epithelial cells before spreading to other cell types, especially B cells; however it also infects lymphocytes, resulting in an atypical morphology referred to as Downey cells. Most infected cells do not produce infectious virions, but select portions of the viral genome. The virus remains latent in B lymphocytes. Over 90% of the population have antibody to against the virus.

The heterophile test (Monospot Test) is used as a primary screening for infectious mononucleosis. The test measures cross-reacting polyclonal IgM antibodies stimulated by the infection of B cells by the causative virus. Other infections, such as hepatitis B virus, may also produce a similar expression of heterophile antibodies. Early in the course of the infection the heterophile test is negative since the immune system has not been adequately stimulated. In addition, a small percentage of the population (approximately 15%) do not produce the response. The heterophile response is short-lived lasting only 6 months. Later in the infection a specific, high titer IgG directed against the viral capsid antigen, is detectable in the patient's serum.

Viral infection of lymphocytes results in a transient suppression of the cell-mediated immune response (CMI). The resolution of the suppression appears to correspond with the recovery of normal lymphocyte number and function. The time required for resolution of the illness may take many months, with recovery of virus from nasal washings lasting in excess of 12 months. Body fluids such as saliva, semen and blood may harbor the virus. Transmission can occur via direct or indirect contact with any of these fluids. Normal transmission though the ingestion of food has not been documented.

Items 193-198

In May, at a roadside diner, a 23 year-old young man complained to a waitress that he had been bitten by a small but surprisingly aggressive raccoon. The man did not seek medical treatment since he consider the bite to be a minor annoyance. A month later the affected hand became weak. A week later the man died. Before slipping into a coma and death, the man suffered from episodes of rigidity and breath holding, hallucinations and difficulty in swallowing liquids. He experienced excessive drooling, high fevers and disorientation.

193. The symptoms exhibited by the young man are consistent with those of

(A) rabies
(B) Eastern equine encephalitis
(C) Coxsackie B encephalitis
(D) herpes encephalitis
(E) measles encephalitis

194. A brain tissue biopsy would show the presence of

(A) Paschen bodies
(B) Negri bodies
(C) elementary bodies
(D) reticulate bodies
(E) Guarnieri bodies

195. The long incubation period associated with infections caused by this virus is due to the

(A) over abundance of defective viruses in the infecting inoculum
(B) localized nature of the infection
(C) slow migration of the virus through peripheral nerves from striated muscle cells
(D) strong cell-mediated immune response induced by the viral antigens early in the infection
(E) establishment of latency

196. Treatment leading to recovery from infection by the virus is dependent on all of the following factors **EXCEPT**:

(A) early administration of hyperimmune globulin
(B) vaccination
(C) immediate and thorough washing of the wound
(D) immediate medical attention
(E) age of the patient

197. Efforts to control the spread of the virus have included all of the following **EXCEPT**:

(A) control of the reservoir population
(B) vaccination of domestic animals
(C) vaccination of wild animals
(D) improvement of vaccine efficiency
(E) replacement of the wild type virus

198. All of the following viruses belong to the same group **EXCEPT**:

(A) Marburg virus
(B) Vesicular stomatitis virus
(C) Ebola virus
(D) Rabies virus
(E) Pseudorabies virus

ANSWERS AND TUTORIAL ON ITEMS 193-198

The answers are: **193-A; 194-B; 195-C; 196-E; 197-E; 198-E**. Rabies virus infections will invariably result in death if the virus advances to the central nervous system. The natural reservoirs for the virus are raccoon, skunk, fox, wolf and bat populations. An urban form of the disease involves transmission primarily through domesticated animals. Both man, and his carnivorous domestic pets may become part of either natural cycle.

The virus normally is introduced into the body through the bite of an infected animal. Airborne infection from breathing the virus-laden air which accumulates in bat cave has been reported. Viruses replicate in the skeletal muscles before crossing the neuromuscular junctions and ascending through the peripheral nerves into the brain. The process is slow due to the indirect migration of viruses to the brain. Viral replication in the brain, especially the hippocampus, cerebral cortex, cerebellum and dorsal spinal ganglia, results in necrosis and paralysis. A biopsy of the brain would show the presence of cytoplasmic inclusions of viral replication, called Negri bodies, in infected cells. The swallowing reflex is affected, resulting in hydrophobia. The virus also replicates in the salivary glands, resulting in production of abundant, virus-laden, frothy saliva.

Early medical treatment and attention is essential in preventing the disease. Measures such as cleaning the wound, injection of hyperimmune globulin, and vaccination contribute to survival. The slowness of viral migration (several weeks in some cases) allows the stimulation of the individual's immune system in combating the disease. New recombinant vaccines ease the discomfort experienced with the old vaccines which required numerous abdominal injections.

Effective measures designed to control the spread of rabies have included attempts to control the reservoir populations through limited harvesting, import bans and mandatory vaccinations for dogs and cats. Recent experiments have proven the feasibility of vaccinating wild animals. Attenuated virus vaccines and synthetic baits make it possible to wild immunize animals in large areas. It is more reasonable to immunize wild animals by protecting them through vaccination than to replace the existing virus with a different virus.

The pseudorabies virus is an icosahedral DNA virus belonging to the herpes virus group. The rabies virus and other rhabdoviruses are helical RNA viruses.

<u>Items 199-205</u>

A 20 year-old college student was hospitalized with chief complaints of severe headache, nausea, vomiting and fever. Physical examination revealed nuchal rigidity. A spinal tap yielded results consistent with the diagnosis of viral meningitis. Bacteriological culture was negative. Careful electron microscopic examination of the CSF sediment showed 200 nm pleomorphic viral particles with projection-studded envelopes. A review of the present illness revealed a brief episode of testicular pain with mild swelling and tenderness.

199. All of the following are true regarding the most likely etiological agent of this disease **EXCEPT**:

(A) There is an available vaccine
(B) The viral genome is itself a messenger RNA
(C) The agent is capable of viremic dissemination in the body
(D) The parotid gland can be its other target organ
(E) The virus agglutinates chicken erythrocytes

200. Which of the following specimens is the least likely to contain significant amount of the causative agent?

(A) Cerebrospinal fluid
(B) Blood
(C) Urine
(D) Saliva
(E) Tears

201. The meningitis observed in the patient is due to infection by the

(A) measles virus
(B) herpes simplex virus
(C) influenza A virus
(D) mumps virus
(E) Coxsackie B virus

202. This virus has how many serotypes?

(A) 1
(B) 2 - 5
(C) 6 - 10
(D) 11 - 20
(E) more than 20

203. A key pathological feature of the disease is

(A) meningitis
(b) orchitis
(C) oophoritis
(D) parotitis
(E) viremia

204. The highest frequency of infections by the above virus is observed in individuals aged

(A) 1 to 5 years old
(B) 5 to 15 years old
(C) 15 to 25 years old
(D) 25 to 45 years old
(E) over 45 years old

205. Transmission of the virus is primarily through

(A) blood and blood products
(B) inhalation of aerosol droplets
(C) ingestion of contaminated foods
(D) sexual contact
(E) the bite of a vector

ANSWERS AND TUTORIAL ON ITEMS 199-205

The answers are: **199-B; 200-E; 201-D; 202-A; 203-D; 204-B; 205-B**. The mumps virus is an enveloped RNA virus. They belong to the paramyxovirus group characterized by prominent hemagglutinin antigens present in the envelope. The genome consists of a single negative-sense RNA molecule rendering it unusable as RNA prior to transcription. The virus is initially acquired by inhalation of contaminated aerosol droplets with the initial rounds of replication taking place in the upper respiratory tract. The virus spreads rapidly, affecting a wide range of tissues as it is disseminated throughout the body during viremic stages. The parotid salivary glands characteristically become greatly inflamed making it difficult for the individual to swallow. The virus may also infect the testes (orchitis), producing either unilateral or bilateral infections. Similarly infections of the ovary (oophoritis) may occur in women.

A significant number of individual experience a meningitis infection during a mumps illness. The response is usually mild but for a small number of individuals the more severe encephalitis may develop. Virus may be recovered from the CSF of patients.

Infections occur most frequently in children between the ages of 5 to 15 years old. Infections in neonates are rare. Infections in young adults (especially men who may become sterile) increases the likelihood of significant complications.

The mumps virus has only one serotype. Therefore, vaccine development was relatively simple. The current vaccine is an effective attenuated vaccine. Natural infections usually result in natural immunity against reinfection.

Items 206-210

A 35 year-old woman suffering from acute lymphocytic leukemia and undergoing chemotherapy suddenly developed small vesicular eruptions across her left breast. The eruptions were asynchronous, painful and lasted 6 days before clearing. The affected area developed small vesicles which progressed to ulcerated and crusted pustules. She had a similar occurrence last year.

206. The appearance of localized skin lesions over a specific dermatome is characteristic of

(A) chickenpox
(B) shingles
(C) generalized herpes simplex
(D) measles
(E) rickettsialpox

207. A cytologic examination of stained smear made from material obtained from the lesions (a Tzanck test) would indicate

(A) epithelial cells with cytoplasmic inclusions
(B) evidence of multinucleated giant cell formation and nuclear inclusions
(C) many lymphocytes with vacuolated cytoplasm
(D) bacterial contamination and cell necrosis
(E) normal cell morphologies

208. The causative agent is most likely the

(A) zoster virus
(B) herpes simplex virus
(C) smallpox virus
(D) measles virus
(D) vaccinia virus

209. The patient had to have been infected with the same virus earlier in life and experienced a case of

(A) chickenpox
(B) German measles
(C) shingles
(D) variola
(E) hemorrhagic fever

210. Transmission of her disease occurred by

(A) inhalation of contaminated aerosol droplets
(B) ingestion of contaminated foods
(C) transfusion of blood or blood products
(D) activation from latency
(E) sexual contact

ANSWERS AND TUTORIAL ON ITEMS 206-210

The answers are: **206-B; 207-B; 208-A; 209-A; 210-D**. The herpes zoster virus causes shingles. It is identical to the varicella virus which causes chickenpox. Exposure to the varicella virus results in the establishment of a latent state in many individuals. The virus may remain dormant for many years before being activated and expressed as a different disease, shingles.

There are numerous factors which have been attributed to the reactivation of the zoster virus. Exposure to sunlight, chemotherapy, immunosuppression, menstrual cycle changes, trauma, extreme mental depression or anxiety are but a few factors reported to induce the expression of shingles.

Shingles affects a specific dermatome. Fluid recovered from the lesions contains active virus capable of infecting unexposed individuals. The patient rarely autoinfects with new dermatomal exposure. It is believed that both humoral and cell-mediated immune responses restrict the spread of the virus to new body sites, but are unable to eradicate the virus from the body. Interferon may stabilize the infection.

There are no effective treatment protocols at present. Hyperimmune globulin is effective in suppressing varicella infections in cases of known exposure. Immunosuppressed individuals are at risk from varicella and therefore would benefit from passive immunization. Vaccine development is progressing slowly.

Items 211-215

An 18 year-old female had an episode of mild infection. Approximately six days later, she developed severe headaches, vomiting and fever. Neurologically, she was oriented and normal, except for a sight degree of nuchal rigidity. A spinal tap revealed CSF high in cell count and protein level. Bacteriological culture of the CSF resulted in no growth. The physician's report indicated aseptic meningitis.

211. The infecting agent responsible for the illness is likely a(an)

(A) rhinovirus
(B) Coxsackie B virus
(C) mumps virus
(D) echovirus
(E) rhabdovirus

212. An electron microscopic analysis of the particles recovered from the CSF would reveal

(A) enveloped viruses with globular projections
(B) naked icosahedral viruses with projections at each vertex
(C) minute naked viruses
(D) bullet-shaped viruses with surrounding envelopes
(E) double-shelled icosahedral viruses

213. The virus was probable acquired by

(A) inhalation
(B) ingestion
(C) transfusion of blood or blood products
(D) activation from latency
(E) sexual contact

214. Infections of this type

(A) usually self-resolve
(B) require drastic antiviral chemotherapy
(C) usually progress into fatal illnesses
(D) recur as latent infections
(E) appear genetically linked

215. Large concentrations of the virus may be found in

(A) blood
(B) CSF
(C) feces
(D) saliva
(E) urine

ANSWERS AND TUTORIAL ON ITEMS 211-215

The answers are: **211-B; 212-C; 213-B; 214-A; 215-C**. The Coxsackie viruses are enteroviruses. They are capable of infecting a wide range of body tissues; however, cells of the gastrointestinal tract are severely infected, resulting in the shedding of many viruses in feces. After the initial infection by ingestion, the viruses replicate in the oral cavity and gastrointestinal tract. Viremic stages transport the virions to distant tissues and, in some cases, into the central nervous system where meningitis or even encephalitis may result. Clinical illnesses last from 1-2 weeks.

Complications such as myocarditis or pericarditis may occur as a result of Coxsackie virus infections. Most infections are subclinical in nature.

There are many serotypes of Coxsackie A and B viruses. Coxsackie B viral infections are most frequently associated with aseptic meningitis and myocarditis. However most infections are self-limiting. Complications such as chronic heart disease and encephalitis may be significant. Currently there is no prophylaxis for enterovirus infections. Treatment is only supportive.

Items 216-220

A pregnant 17 year-old inner city female developed a very faint macular rash over a portion of her face, neck and upper back about one month into her pregnancy. The young woman only noticed a mild fever and symptoms resembling the "flu" along with some minor stiffness in her hands and knees. Her pregnancy was uneventful even though she received no prenatal care. The baby was born three weeks earlier that anticipated. A few days after delivery, it became apparent that the newborn did not respond to sound.

216. Umbilical cord blood would most likely show antibody against

(A) herpes simplex virus
(B) cytomegalovirus
(C) Epstein-Barr virus
(D) rubella virus
(E) rubeola virus

217. Congenital infection occurs as a result of

(A) viral toxins crossing the placenta
(B) viral toxins impairing maternal nourishment of the fetus
(C) transplacental spread of virus and growth in the fetus
(D) autoimmune response of maternal cells against fetal tissues
(E) autoimmune response of fetus against maternal antigen

218. At three months, the child's serum will most likely show

(A) only fetal IgM against the virus
(B) only fetal IgG against the virus
(C) fetal IgM and IgG along with maternal IgG against the virus
(D) fetal IgM and IgG along with maternal IgM and IgG all directed against the virus
(E) only maternal IgG against the virus

219. The infection is usually acquired through

(A) ingestion of contaminated foods
(B) inhalation of aerosol droplets
(C) the bite of an arthropod vector
(D) contaminated blood products
(E) sexual contact

220. Infection of newborns or young children by the same virus usually results in

(A) mild infections with few obvious symptoms
(B) significant neurological damage
(C) transient neurological damage
(D) significant cardiac damage
(E) minor cardiac damage

ANSWERS AND TUTORIAL ON ITEMS 216-220

The answers are: **216-D; 217-C; 218-C; 219-B; 220-A**. Rubella (German measles, 3-day measles) is caused by rubella virus, a kind of togavirus. The infections occur most frequently during the winter and spring months with symptoms, in many cases, closely resembling those of influenza. The infection is usually mild to inapparent in young children and adults, however in pregnant women the outcome of an infection could be major. The rash, which develops about 2 weeks after infection, distinguishes it from many other viral infections. German measles infections during first trimester are especially critical because this is a developmental period where organogenesis is initiated. Developing organs, especially the brain and sense organs, are particularly sensitive to rubella virus-induced teratogenesis. Viral transmission occurs primarily through the inhalation of aerosol droplets. The virus undergoes a viremic stage in the infected mother enabling transplacental infection of the developing fetus. The virus may replicate in a wide range of fetal tissues thereby affecting a wide range of congenital defects of including mental retardation, blindness, deafness, anemia, hepatitis, and stillbirths.

Surviving newborns, up to 5 months of age, will usually possess residual maternal IgG against the rubella virus along with fetal IgM and IgG. Infected infants will continue to shed the virus in their urine and oral cavity for extended periods of over 12 months. They act as reservoirs for the virus and can infect other youth and unvaccinated young adults.

There are no specific treatments for the infection. Containment is best accomplished through vaccination programs. The current attenuated vaccines are imperfect. Unfortunately, they can cause infections in pregnant women. Nonpregnant women should not take the vaccine if they expect to conceive within the next 3 months.

Items 221-226

(A) Anti-HBe
(B) Anti-HBc
(C) Anti-HBs
(D) Anti-HBpol
(E) HBeAg
(F) HBcAg
(G) HBsAg
(H) HBpolAg
(I) HBV
(J) DNA
(K) RNA
(L) DNA\RNA

For each of circumstance in the items below, match the most likely component of immune response against a hepatitis B virus infection in the answers above.

221. An IgG response necessary for the neutralization of the virus.

222. An early IgM response to a viral enzyme.

223. The antigen used in the current recombinant vaccine.

224. An IgG response not found in the serum of chronic infections.

225. The infectious virus.

226. The type of nucleic acid found to the capsid of the infectious virus.

ANSWERS AND TUTORIAL ON ITEMS 221-226

The answers are: **221-C; 222-D; 223-G; 224-C; 225-I; 226-J**. The hepatitis B virus is a small icosahedral virus surrounded by an envelope consisting of the surface antigen (S-antigen), which may exist in any of six different forms. The S-antigen is the neutralizing antigen, therefore the production of serum IgG against this antigen (in the form of Anti-HBs) is essential for recovery from infection. An early IgM response against the core and surface antigens is not sufficient for neutralization. An early stage in the disease, termed the "window phase", is marked by an absence of HBsAg and anti-HBs in the blood. Anti-HBc is present and used as an indicator of infection. The current recombinant vaccine is derived from the S-antigen produced in yeasts, purified and used as a vaccine antigen. The old human blood plasma-derived vaccine is no longer used in the U.S.

The immune response to infection is directed against several antigens. Similarly the genome of the virus has several unusual configurations. Anti-HBe (A) is a response directed against this different core antigen. This antibody indicates low transmissibility of infection.

Anti-HBc (B) is directed against the core protein. The IgM class is short lived and indicative of a recent infection. The IgG is longer lasting. It is present during the "window phase" of the disease and used as an indicator of infection. Anti-HBs (C) is the major neutralizing and protective antibody response against HBV. Anti-HBpol (D) is a response directed against the virus polymerase. Although it appears early in infection, it is not widely used diagnostically.

HBeAg (E) is an immature core antigen. Its presence indicates low transmissibility. HBcAg (F) is the mature core (capsid) antigen. It is associated with infectious virions. HBsAg (G) is the surface or envelope antigen of the virus. It is abundant in carriers or acutely infected individuals prior to the expression of anti-HBs. HBpolAg (H) is the replicating polymerase of the virus. It directs both DNA and RNA synthesis using the different virus genome stages (RNA and DNA\RNA) as templates. HBV (I) is the infectious virion, also called the "Dane particle". It is the infectious stage of the virus consisting of genome, core and surface envelope antigens.

DNA (J) is the stage of the virus genome. It consists of an incomplete double-stranded DNA circle. RNA (K) is a transitional stage of viral replication. The virus genome goes through a single-stranded RNA stage during replication. DNA\RNA (L) is another intermediate stage of the HBV genome. The final double-stranded DNA genome is copied off this DNA\RNA intermediate stage.

Items 227-231

Two working mothers left 9 month-old infants at a day care center. One Thursday evening, both children experienced acute gastroenteritis with fever, vomiting and watery diarrhea. Their doctors prescribed fluids and rest until the condition resolved itself or worsened. Test results eliminated bacterial infection and food poisoning as a disease source.

227. A virus known to cause diarrhea among newborns and young children are the

(A) coronaviruses
(B) adenoviruses
(C) rhinoviruses
(D) polioviruses
(E) rotaviruses

228. An appropriate specimen useful in the diagnosis of the above virus is a

(A) blood sample
(B) nasal swab
(C) stool sample
(D) nasal washing
(E) CSF sample

229. The viruses are transmitted between children primarily by

(A) inhalation of aerosol droplets
(B) blood or blood products
(C) ingestion of or oral contact with contaminated items
(D) pet scratches
(E) bite of an arthropod vector

230. The viruses localize primarily in the

(A) esophagus
(B) stomach
(C) duodenum and proximal jejunum
(D) large intestine
(E) entire gastrointestinal tract

231. Breast-fed infants may benefit from the presence of the following immunoglobulin against the virus

(A) IgG
(B) IgA
(C) IgM
(D) IgE
(E) IgD

ANSWERS AND TUTORIAL ON ITEMS 227-231

The answers are: **227-E; 228-C; 229-C; 230-C; 231-B**. Viral gastroenteritis in infants between 6 and 24 months is most often due to rotavirus infections. There are 4-5 serotypes of the virus. Infections arise from ingestion of the virus from contaminated objects. After 1-3 days of incubation, with growth of the virus in the gastrointestinal tract, the symptoms appear for the next week.

Infection of the duodenum and proximal jejunum cause destruction of the epithelial cells and mucosa, leading to fluid loss. Large quantities of the virus are recoverable from the stool. They are detectable by electron immunoprecipitation, enzyme-linked immunosorbent assay and radioimmunoassay. Electron microscopy reveals double-shelled viruses. Additional test will show a double-stranded RNA genome, different from other RNA animal viruses. By age 5, most children have IgG against the viruses. Children born of previously exposed mothers may receive maternal IgA in colostrum or breast milk.

Items 232-238

(A) Ribavirin
(B) Interferon
(C) Dideoxyinosine
(D) Methisazone
(E) Acyclovir
(F) Amantadine
(G) Gancyclovir
(H) Iododeoxyuridine
(I) Azidothymidine
(J) Phosphonoacetic acid
(K) Anti-sense message RNA

For each antiviral compound above, match the most appropriate statement describing its mode of action in the items below.

232. Blocks the assembly of poxviruses by inhibiting viral protein synthesis.

233. Nucleoside analogue that causes DNA chain termination, used as a back-up agent in the treatment of AIDS.

234. Small compound which binds to the DNA polymerase of the herpes viruses.

235. Anti-herpes and anti-varicella-zoster compound, phosphorylated by viral kinases to monophosphates, later to triphosphates by cellular enzymes, eventually blocking viral DNA polymerase.

236. Interferes with the influenza viral attachment to and uncoating by host cells.

237. Host species-specific antiviral compound effective against a wide range of viruses.

238. New theoretical approach for indirectly inhibiting HIV by blocking the synthesis of transactivator proteins.

ANSWERS AND TUTORIAL ON ITEMS 232-238

The answers are: **232-D; 233-C; 234-J; 235-E; 236-A; 237-B; 238-K**. The effectiveness of any antiviral compound is dependent upon several variables. These include: selective inhibition of viral attachment, replication and maturation; effectiveness at physiological conditions; non-toxicity to host cell functions and nonmutagenicity. Many antivirals are nucleoside analogues designed empirically to induce mutations or interfere with DNA synthesis.

Ribavirin (A) is active against both DNA and RNA viruses. It inhibits guanosine 5' phosphate and has been used in treating HIV.

Interferons (B) inhibit and degrade the viral mRNA of a broad range of viruses. They also block the elongation of newly synthesized viral mRNAs and act through the regulation of host cell factors.

Dideoxyinosine (C) is a nucleoside analogue that causes DNA synthesis termination. It has been used in the treatment of HIV.

Methisazone (D) blocks the synthesis of mRNA and proteins and consequently the assembly of the poxviruses. It has been used in treating smallpox.

Acyclovir (E) is a DNA chain terminator. It is phosphorylated by viral enzymes to the monophosphate and further by host enzymes before incorporation into DNA. It has been used to treat herpes simplex encephalitis.

Amantadine (F) prevents the attachment and uncoating of the influenza A virus to host cells. It has no effects against influenza B and C viruses.

Gancyclovir (G) is a nucleoside analogue which has a four-carbon open chain sugar replacing ribose. The structure closely resembles acyclovir and is used to treat cytomegalovirus infections.

Iododeoxyuridine (H) is a halogenated thymidine analogue. The phosphorylated compound is incorporated into DNA resulting in incorrect base insertions. It has been used in treating herpes simplex eye infections.

Azidothymidine (I) is a nucleoside analogue that causes DNA chain termination. It is most effective against viruses with reverse transcriptase activities such as HIV.

Phosphonoacetic acid (J) is a small molecule which is a potent and specific inhibitor of DNA polymerase in herpesviruses. It has been proposed as a new treatment against herpes simplex virus; and, although effective, it has significant nephrotoxicity.

Anti-sense message RNA (K) consists of the complementary nucleotide sequences of a particular viral RNA. The base pairing would theoretically inhibit the mRNA translation. Anti-sense sequences against the mRNA HIV regulatory proteins such as tat and rev have been experimentally tested with promising results.

Items 239-244

A cautious office worker decided to get vaccinated against the seasonal (1992-93) flu outbreak. Her physician vaccinated her with an intramuscular dose containing:

A/Texas/36/91 (H1N1)
A/Beijing/353/89 (H3N2)
B/Panama/43/90

Four weeks after receiving the vaccine, the young woman developed a high fever, myalgia, headaches, chills and a dry unproductive cough. The cough became progressive with the general symptoms lasting 3 weeks. Her doctor told her that she had the flu even though serological tests showed that the woman seroconverted.

239. The patient's serum would most likely be capable of partially neutralizing all of the following virus isolates **EXCEPT**:

(A) A/Texas/36/91 (H1N1)
(B) A/Newark/53/89 (H3N1)
(C) A/Washington/43/91 (H1N2)
(D) A/Beijing/353/89 (H3N2)
(E) A/Beijing/111/89 (H2N3)

240. Small changes in the surface glycosylation of the influenza virus are referred to as

(A) genetic shift
(B) genetic drift
(C) camouflage
(D) genetic reassortment
(E) point mutations

241. The primary neutralizing antigen of the Influenza A virus is the

(A) viral polymerase
(B) neuraminidase
(C) hemagglutinin
(D) matrix protein
(E) nuclear protein

242. Children who experience progressive influenza illness may develop liver and cerebral edema called

(A) Reye's syndrome
(B) Guillain-Barré syndrome
(C) egg drop syndrome
(D) Down syndrome
(E) Löffler's syndrome

243. A major concern regarding influenza infections in the elderly is

(A) immune suppression caused by the virus
(B) the viremic spread of the virus to the central nervous system
(C) secondary bacterial infections
(D) drug resistance in the infecting virus
(E) reversion to enhanced virulence

244. Which of the following specimen sources is best suited for the recovery of live influenza viruses?

(A) blood
(B) urine
(C) CSF
(D) nasopharyngeal swabs
(E) feces

ANSWERS AND TUTORIAL ON ITEMS 239-244

The answers are: **239-E; 240-C; 241-C; 242-A; 243-C; 244-D**. Seasonal outbreaks of the influenza viruses occur each year in the U.S. and worldwide. Humans, birds and swine may harbor the same virus strains thus providing opportunities for genetic reassortment to occur. The genome of the viruses consists of eight RNA segments, each encoding for a specific viral product, such as the hemagglutinin or neuraminidase. Genetic reassortment, also called genetic shift, may result in the emergence of a new serotype of the virus unknown to the population. Pandemics and epidemics result from such changes in the virus antigenicity within a certain host range. Vaccinations are designed to provide immunity against the most likely encountered serotype of the hemagglutinin, and to a lesser extent the neuraminidase. There are roughly 13 different recognized hemagglutinin designations and 9 neuraminidase designations. Each represented by H or N, and a number, respectively. Vaccine containing antigen derived from a H1 isolate will not protect against an H3 isolate.

Small point mutations in the envelope antigens of the virus, referred to as genetic drift, are less dramatic. Such changes unusually only result in minor antigenic variations which remain cross reactive with parent isolates. "Camouflage" are changes in the extent of glycosylation of surface glycoproteins which may alter the antigenic presentation of the virus in a particular host. These enzymatic variations may change between hosts.

Influenza infections in children who have ingested salicylate have also been linked to Reye's syndrome, an illness characterized by vomiting, lethargy and various CNS symptoms. Other viral infections, such as herpes simplex infections, may produce a similar complication. A primary concern, especially regarding influenza in the elderly, is the complications associated with secondary bacterial infections. Streptococcus pneumoniae, Haemophilus influenza, and Staphylococcus aureus infections present the most concern. Fatalities run high in this high risk population.

The influenza viruses do not present the typical viremic stage. Viruses, for isolation, are best recovered from nasopharyngeal or throat swabs. Inhaled viruses primarily infect the respiratory tract resulting in ciliary dysfunction and epithelial necrosis. Infiltration of monocytes also occurs. Resolution of the infection usually occurs after 2 - 10 weeks along with the development of a high serum IgG response to the neutralizing (hemagglutinin) antigen.

Items 245-248

A working mother left her young infant at a nursery each work day. On Monday, she noticed two other children at the nursery with runny noses and coughs. By Friday, her daughter developed similar symptoms and breathing difficulty. A physical examination by the pediatrician indicated bronchiolitis with bilateral interstitial infiltrates and hyperexpansion of the lungs. The mother was deeply concerned. The physician recommended fluids and rest for the infant since the infection appeared viral with no apparent complicating bacterial infections.

245. The virus most likely responsible for the infection is a

(A) respiratory syncytial virus
(B) influenza virus
(C) parainfluenza virus
(D) coronavirus
(E) rhinovirus

246. The causative virus contains all of the following **EXCEPT**:

(A) an RNA genome
(B) viral nucleoprotein
(C) matrix protein in the envelope
(D) hemagglutinin in the envelope
(E) glycoprotein in the envelope

247. The single most important viruses responsible for infant respiratory diseases are the

(A) respiratory syncytial viruses
(B) influenza viruses
(C) parainfluenza viruses
(D) coronaviruses
(E) rhinoviruses

248. Exfoliated cells, recovered from a nasopharyngeal swap, stained with a monoclonal fluorescent antisera against the viral capsid proteins would show, under fluorescent microscopic analysis

(A) some multinucleated cells with nuclear fluorescence
(B) single cells with nuclear fluorescence
(C) single and multinucleated cells with cytoplasmic fluorescence
(D) single and multinucleated cells with nuclear fluorescence
(E) multinucleated cells with both nuclear and cytoplasmic fluorescence

ANSWERS AND TUTORIAL ON ITEMS 245-248

The answers are: **245-A; 246-D; 247-A; 248-C**. Respiratory syncytial viruses (RSV) are the most prevalent cause of lower respiratory tract disease (such as pneumonia and bronchiolitis) in infants. Fast-spreading outbreaks occur annually during the winter months. The infection may resemble the common cold, especially in older infants and adults.

The viruses belong to the paramyxovirus group. A major distinguishing feature of RSV is the absence of either hemagglutinin and neuraminidase activity in the viral envelope. Both features are found in other paramyxoviruses. Infected cells may fuse with adjacent cells forming large syncytia which contain cytoplasmic inclusions of assembling viruses. Fluorescent microscopic analysis of infected exfoliated cells or of infected cell cultures, using antibody directed against the capsid protein, will show cytoplasmic fluorescence. Serological tests of a 4-fold, or greater, rise in serum antibody titer (against RSV) is also diagnostic. Even though only one serotype of the virus exists there is no widely used vaccine. Ribavirin has been recommended for the treatment of severely ill infants.

Items 249-254

A small group of soldiers, deployed to a remote Somalian village during the Winter of 1992, encountered a young female patient who had large (1 cm) pustules over her face, arms and palms. There were also a few on her stomach. All pustules were about the same size and at about the same developmental stage. Her serum did not react with a latex agglutination assay designed to detect varicella and herpes simplex viruses.

249. The appearance and location of the pustules on the young woman are consistent with a presumptive diagnosis of

(A) chickenpox
(B) herpes simplex infection
(C) measles
(D) smallpox
(E) an arbovirus infection

250. Exfoliated cells, present in vesicle fluid, would show

(A) multinucleated cells with nuclear inclusions
(B) multinucleated cells with cytoplasmic inclusions
(C) primarily single cells with nuclear inclusions
(D) primarily single cells with cytoplasmic inclusions
(E) cells with both nuclear and cytoplasmic inclusions

251. The physical characteristics of the causative virus consists of the following:

(A) double-stranded RNA in segments and no envelope
(B) single-stranded RNA and envelope with hemagglutinin
(C) double-stranded DNA and envelope with hemagglutinin
(D) double-stranded DNA and no envelope
(E) single-stranded DNA and no envelope

252. The virus is transmitted primarily by

(A) inhalation of aerosol droplets
(B) ingestion of contaminated foods
(C) injection of contaminated blood or blood fluids
(D) animal scratches
(E) vector bite

253. This rare disease could easily be misdiagnosed as

(A) chickenpox
(B) herpes simplex infection
(C) measles
(D) smallpox
(E) an arbovirus infection

254. An attenuated vaccine used in the prevention of the illness uses the

(A) measles virus
(B) vaccinia virus
(C) varicella virus
(D) yellow fever virus
(E) mumps virus

ANSWERS AND TUTORIAL ON ITEMS 249-254

The answers are: **249-D; 250-D; 251-C; 252-A; 253-A; 254-B**. The WHO declared that smallpox (variola major and variola minor) had been eradicated in 1976. The last reported naturally-occurring case occurred in the early 1970's in Somalia. Though highly unlikely, new cases could occur in isolated regions lacking public health measures.

The virus is acquired by inhalation of aerosol droplets followed by initial rounds of replication in the upper respiratory tract. Infected cells show the characteristic cytoplasmic inclusions. Infected cells lyse and release large, double-stranded, enveloped, DNA viruses into the blood stream. Dissemination and spread of the virus to all internal organs coincided with the appearance of pustular skin lesions. The synchronous nature of the lesions could be misdiagnosed as chickenpox although lesions appear asynchronous in chickenpox. Fatalities from the infection by virulent strains approach 20%.

The vaccine, using attenuated vaccinia virus, presented a large number of complications due to its contamination with bacteria causing generalized infections and encephalitis. Immunocompromised patients were especially prone to these side effects. Methisazone, an inhibitor of poxvirus assembly, and rifampin, an inhibitor of poxvirus DNA-dependent RNA polymerase, have been used to treated patients.

Items 255-258

Fifteen men, all in their late twenties, developed progressive neurological disorders. Symptoms included presenile dementia and ataxia. Serum and CSF antibody were negative in ELISA tests against a battery of known neurotropic viruses. Each of the men had received, 15 years earlier, a series of treatments with pituitary growth hormone. A preliminary diagnosis of Creutzfeldt-Jakob disease was made.

255. The causative agent of the illnesses has been termed a

(A) slow virus
(B) unconventional agent
(C) classical virus
(D) latent virus
(E) virus

256. A similar disease affecting sheep or cattle is called

(A) kuru
(B) scrapie
(C) mink encephalopathy
(D) visna
(E) progressive multifocal leukoencephalopathy

257. All of the following descriptions of the causative agent or infection are correct **EXCEPT**:

(A) Infection stimulates no humoral immune response
(B) The incubation period may exceed 20 years
(C) Irregular host proteins cause the disease
(D) The infectious agent is spread by inhalation
(E) The infectious agent resists strong aldehydes.

258. The prion protein has a characteristic MW =

(A) 10,000 - 20,000
(B) 20,000 - 30,000
(C) 30,000 - 40,000
(D) 40,000 - 50,000
(E) 50,000 - 60,000

ANSWERS AND TUTORIAL ON ITEMS 255-258

The answers are: **255-B; 256-B; 257-D; 258-B**. Creutzfeldt-Jakob disease is caused by an unconventional agent (virus). The infectious agent, called the prion, consists of aberrant forms of a normally-occurring host cell protein. The altered protein has the ability to "self-replicate" and accumulate in cells, resulting in cell necrosis. The pathology is localized primarily in the brain and develops over 15-20 years or more.

The present case history resulted from the accidental inclusion of infectious material, believed derived from the contaminated pituitary glands used in the extraction of human growth hormone. Hormone produced by recombinant DNA technology eliminates the potential of similar contaminations.

The infections agent, prion protein, is a glycosylated protein of MW $\approx$27,000. Diseased individuals do not produce an immune response to the prion protein. A characteristic of the protein is its resistance to factors which readily inactivate conventional viruses (i.e. formaldehyde, heat, nucleases, proteinases, lipid solvents and irradiation).

Scrapie, the prototype unconventional agent, is a prion-directed disease of cattle and sheep. Kuru was the first known human prion-induced disease. There is no treatment or vaccine available for stemming the spread of the disease. Natural transmission of the disease occurs through the ingestion of contaminated foods or exposure to contaminated materials.

Items 259-261

(A) Western equine encephalitis
(B) St. Louis encephalitis
(C) Yellow fever
(D) California encephalitis
(E) Sandfly fever
(F) Colorado tick fever
(G) Eastern equine encephalitis
(H) Rift valley fever
(I) Dengue fever
(J) Venezuelan equine encephalitis

For each of the following descriptions, match the most likely arbovirus disease occurring in the U.S.

259. Most common bunyavirus disease, animal reservoir is small mammals.

260. Most common arboviral diseases.

261. Most common alphaviral disease.

Items 262-265

There are over 400 different arthropod borne (arbo) viruses causing disease. Arboviruses are grouped based on the physical properties of the infectious virions and their mechanism of maturation in the host cell.

262. Arboviruses which are transmitted by mosquitoes are harbored in the salivary glands of the insect. This feature

(A) insures likely transmission at each feeding event
(B) becomes lethal to the vector
(C) affects both sexes of the mosquito population
(D) produces a long viremic stage in the subsequent host animal
(E) requires only 1-2 bites for adequate infecting dose accumulation

263. Urban cycles of arboviral infections refer to infectious cycles in which

(A) animal reservoirs are small mammals
(B) animal reservoirs are birds
(C) reservoirs are humans
(D) humans are the accidental hosts
(E) either animals or humans are the reservoirs

264. Classical "breakback fever" is an arboviral infection not frequently encountered in the U.S. It associated with severe joint and muscle pain, maculopapular rash, cough and high fever. The causative agent is the

(A) sandfly fever virus
(B) dengue fever virus
(C) yellow fever virus
(D) Venezuelan equine encephalitis virus
(E) California encephalitis virus

265. An early effective human vaccine has been used in the prevention of

(A) Western equine encephalitis
(B) Eastern equine encephalitis
(C) yellow fever
(D) Colorado tick fever
(E) dengue fever

ANSWERS AND TUTORIAL ON ITEMS 259-265

The answers are: **259-D; 260-F; 261-A; 262-A; 263-C; 264-B; 265-C**. There are over 400 different viruses which require an arthropod vector in the transmission of disease. Collectively they are referred to as arboviruses (for arthropod-borne viruses). Most arboviruses are either togaviruses or bunyaviruses. Togaviruses, consisting of alphaviruses (60-70 nm) and flaviviruses (45-55 nm), are small, positive polarity, single-stranded RNA viruses. Bunyaviruses are larger (90-100 nm) and have segmented, negative polarity RNA genomes. Arboviruses also include other viral groups such as the reoviruses and rhabdoviruses.

Relatively few arboviral infections occur annually in the U.S. Colorado tick fever virus, a reovirus, causes the single most prevalent infections (200-300/year). California encephalitis virus is the most frequently seen bunyavirus infection. Western equine encephalitis is the most frequently seen alphavirus infection.

Viruses transmitted by mosquitoes and ticks are harbored in the salivary grands of the vector. A period of 1-2 weeks are usually required for the virus to replicate and amplify within the vector in order to achieve a concentration sufficient to cause disease. Birds and small mammals serve as the natural hosts, and reservoirs, for many cycles of infection in the tropical rain forests. Humans are the accidental victims of this type of zoonotic infection. In contrast, humans are the primary reservoir in the urban cycles of infections. A model most extensively studied is that of yellow fever which has well-defined jungle and urban cycles of transmission.

Some arbovirus infections, such as dengue fever, have several forms with markedly different symptoms. The classical dengue fever is also called "breakback fever" due to the high fever and extreme myalgia/arthralgia caused by the disease. Fatalities are relatively rare in contrast to the hemorrhagic fever version of dengue fever, which is a far more severe form of the disease with high (10%) fatalities.

Vaccination, along with vigorous vector control, has been successful in the prevention and containment of arboviral diseases. Most vaccines have been developed for animals. The yellow fever vaccine is a widely used and successful human vaccine. The entire nucleotide sequence of the yellow fever virus has been determined for different strains of the attenuated and wild type virus.

Items 266-270

(A) Human papilloma virus
(B) Hepatitis B virus
(C) Hepatitis C virus
(D) Herpes simplex virus type 2
(E) Epstein-Barr virus
(F) Human T cell leukemia virus
(G) Human immunodeficiency virus
(H) Rous sarcoma virus

For each of the following human tumors, choose the virus thought to cause the malignant disease.

266. Cervical carcinoma

267. Hepatocellular carcinoma

268. Burkitt's lymphoma

269. Nasopharyngeal carcinoma

270. Hairy-cell leukemia

Items 271-274

Fifty individuals developed similar illnesses within a 1 week interval. All experienced fatigue, poor appetite, fever, nausea, jaundice and dark urine. Medical histories revealed that all had attended the same business convention banquet 4 weeks prior. Bacterial cultures of blood and fecal samples showed no common feature. Symptoms were consistent with hepatitis B infection but anti-HBsAg IgM was negative in all cases.

271. The fact that many of the affected individuals acquired a jaundice-causing illness in conjunction with food consumption indicates possible infection caused by

(A) hepatitis A virus
(B) hepatitis B virus
(C) hepatitis C virus
(D) hepatitis D virus
(E) hepatitis E virus

272. An appropriate serological diagnostic test would look for the presence of

(A) anti-HBe IgG
(B) anti-HAc IgG
(C) anti-HAc IgM
(D) anti-HCc IgM
(E) anti-HEc IgM

273. Infection of those who attended the business dinner were most likely caused by

(A) inhalation of aerosol droplets
(B) contact with contaminated rest room facilities
(C) ingestion of contaminated foods
(D) insect bites
(E) touching small cuts on the hands or scalps

274. A new attenuated vaccine used to prevent the infection releases large numbers of the virus in

(A) tears
(B) saliva
(C) urine
(D) feces
(E) CSF

ANSWERS AND TUTORIAL ON ITEMS 266-270

The answers are: **266-A; 267-B; 268-E; 269-E; 270-F**. The relationship between cancer and tumor viruses in lower animals was established long before similar cancer\virus relationships were established in humans. The presence of several human viruses showed high correlations with certain cancers but causal relationships were difficult to prove. This was due, in part, because of the ethical difficulty of testing Koch's postulates using human subjects. Several causal relationships have been established indirectly.

Human papilloma viruses (A), known to cause benign warts, also cause human cancer. One form of cervical carcinoma is caused by the human papilloma virus 16. DNA sequences from the virus have been identified in certain cervical cancers. A similar relationship has been established between DNA sequences of the hepatitis B virus (B) and hepatocellular carcinoma. The hepatitis B virus shows a preference for human hepatocytes in culture. Individuals suffering from Burkitt's lymphoma and certain nasopharyngeal carcinomas show the presence of DNA sequence homology with the Epstein-Barr virus (E). Certain hairy cell leukemias show the presence of DNA sequences of the human T cell leukemia virus (HTLV-1) (F).

A feature common to all cancers is the integration of the viral genomic DNA into the host chromosome. RNA tumor viruses have reverse transcriptase activity which converts the single-stranded RNA into a double-stranded DNA circular form enabling it to insert into the host chromosome and affect cell transformation. The mechanism of cell transformation is complex and requires multiple events involving viral products altering cellular regulatory factors, many of which have DNA-binding properties.

ANSWERS AND TUTORIAL ON ITEMS 271-274

The answers are: **271-A; 272-C; 273-C; 274-D**. Focal outbreaks of infections involving liver damage may arise from hepatitis A virus. The symptoms of hepatitis A and hepatitis B are almost indistinguishable. Medical histories of the infected individuals will usually identify a presumptive agent. Hepatitis A is usually acquired from the ingestion of contaminated shellfish from polluted waters or foods prepared by a hepatitis A virus carrier. The latter route of infection probably applies in this clinical scenario. A positive serological assay for anti-HAc IgM, antibody against the virus core antigen, arises early in the infection. The latter is detectable especially after jaundice has developed. A serum IgG response develops later in the infection.

The hepatitis A virus is a member of the picornavirus family and therefore has some properties similar to other enteroviruses such as the polioviruses. There is a viremic stage, but much larger numbers of virus are shed in the feces. Poor personal hygiene practices of food handlers, such as not thoroughly washing hands after bowel movements, spread the disease.

There is only one serotype of the hepatitis A virus. A new attenuated vaccine has been designed to prevent the disease and immunize high risk individuals. Passive immunization with immune gamma globulin is effective in preventing disease in probable contacts and decreases the affects of disease in new infections.

Item 275-279

Many animal viruses are composed of different structures that are viral antigens. Match the descriptions of the following viral components in the items below with the most appropriate labeled structure in Figure 2.1 below.

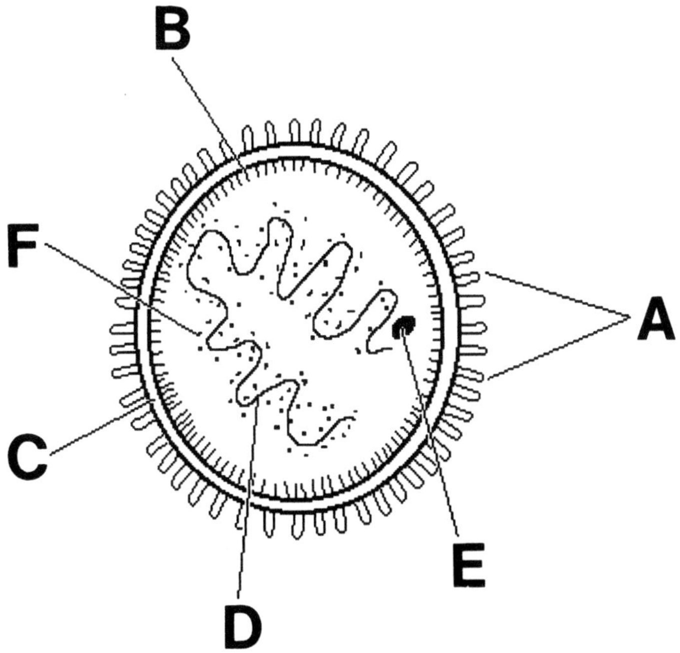

Figure 2.1

275. The host cell is responsible for formation of this component.

276. This antigen is involved in the formation and stabilization of the viral envelope.

277. Infectivity and host range specificity are the functional roles of these antigens.

278. Genetic information is maintained in these molecules.

279. Serotype determinations are based on this set of antigens.

ANSWERS AND TUTORIAL ON ITEMS 275-279

The answers are: **275-C; 276-B; 277-A; 278-D; 279-A**. Enveloped viruses, such as influenza virus, acquire a lipid layer (C) from a host cell membrane and modify it into the viral envelope. The lipid of the influenza virus envelope is derived from the host cell plasma membrane. Other viruses use the nuclear membrane (e.g. herpes viruses), or cytoplasmic membranes (e.g. alphaviruses, and flaviviruses). Unique virally-encoded proteins (A), called peplomers, are inserted into the envelope. Hemagglutinin and neuraminidase are the peplomers produced by influenza viruses. Hemagglutinin is responsible for viral attachment. Neuraminidase functions in the release of mature viruses from infected cells.

The matrix protein (B) functions in the maturation and envelopment of the immature virion and also in stabilizing the newly-formed envelope.

The nucleic acid (D) of influenza consists of eight segments of single-stranded RNA. Each segment encodes a specific viral protein. The RNA is surrounded by a nucleoprotein (F) which also protects it from enzymatic digestion.

The final infectious virion contains a small number of ready-made RNA-dependent RNA polymerase molecules (E). These recognize the viral genome as template for the production of messenger RNA.

Items 280-285

Viral diseases produce symptoms used in preliminary diagnosis. Match the following symptoms with the listed specific viruses and groups of viruses.

(A) Adenoviruses
(B) Influenza viruses
(C) Herpes simplex viruses
(D) Epstein-Barr virus
(E) Mumps virus
(F) Measles virus
(G) Rotaviruses
(H) Reoviruses
(I) Parvoviruses
(J) Vaccinia virus
(K) Coronaviruses
(L) Rhabdoviruses
(M) Togaviruses
(N) Bunyaviruses
(O) Polioviruses
(P) Coxsackie viruses
(Q) Hepatitis A virus
(R) Hepatitis B virus
(S) Hepatitis C virus
(T) Cytomegaloviruses

280. Upper respiratory infection, occurring more frequently during winter months, the common cold is caused by many serotypes of this virus group.

281. Latent infection of body lesions resulting in vesicle formation over the same location at each re-emergence.

282. Primary cause of aseptic meningitis, but also responsible for paralysis and cardiac involvement.

283. Arthropod-borne viral infections, many involving horses, which may cause encephalitis and liver infections in humans.

284. Infection of pregnant women can result in teratogenesis in the developing fetus. Fetal abnormalities include mental retardation, blindness and hearing impairment.

285. Major cause of infant diarrhea.

ANSWERS AND TUTORIAL ON ITEMS 280-285

The answers are: **280-K; 281-C; 282-P; 283-M; 284-T; 285-G**. Most humans have been exposed to a herpes simplex virus (C) as demonstrated by the presence of circulating antibody. The virus remains latent in neuronal cells; and, when activated, results in the appearance of the characteristic skin lesions in the dermatomes served by infected dorsal root ganglia. Many factors such as sunlight, trauma, immunosuppressive drugs and hormonal changes can stimulate the re-expression to the virus. The presence of circulating antibody does not prevent the re-emergence of herpes virus expression.

Rotaviruses (G) are a major cause of infant diarrhea or acute gastroenteritis. The virus is spread by the fecal-oral route and therefore quickly spreads through the children in day-care centers and elementary schools. Children experience diarrhea, vomiting and fever. Vaccination is feasible; however, strict hygiene practices such as hand washing and disposal of fecal wastes aid in the control of epidemics.

Coronaviruses (K) are among the many viruses which cause the common cold or lower respiratory illnesses in adults and children. The glycoproteins of the envelope have a pebble-like appearance. The multiple serotypes of the virus preclude the formulation of an all-inclusive vaccine.

Most of the togaviruses (M) are spread by arthropod (primarily mosquito) vectors. Humans are usually not the natural reservoir. Rather, small mammals, birds, horses and cattle commonly harbor these viruses. Viruses such as the rubella virus, yellow fever virus, Eastern equine encephalitis virus and St. Louis encephalitis virus are representative members of this large collection of exotic disease-causing agents.

Coxsackie viruses (P) are divided into two groups (A and B). They are enteroviruses which may cause debilitating diseases such as aseptic meningitis or myocarditis. Viremia may progress to CNS involvement with virions in the CSF and white cell infiltration. The virus replicates in the lymphoid tissue and epithelial cells of respiratory and gastrointestinal tract.

Cytomegaloviruses (T) (CMV) infect a majority of the population resulting in asymptomatic infections. Pregnant women who lack circulating antibody against CMV may infect their unborn fetus by transplacental infection; or, if the mother is shedding, during the birth process. CMV is known to cause congenital birth defects in infected fetuses. CMV infects a wide variety of tissues and does not show the latency properties of other herpes virus group members.

Items 286-291

The classification of animal viruses is based on the physical properties of the isolated virion. Match the following physical and chemical properties with the listed viruses.

(A) Poxvirus
(B) Herpesvirus
(C) Papovavirus
(D) Parvovirus
(E) Picornavirus
(F) Reovirus
(G) Flavivirus
(H) Togavirus
(I) Orthomyxovirus
(J) Paramyxovirus
(K) Arenavirus
(L) Retrovirus
(M) Rhabdovirus
(N) Bunyavirus
(O) Adenovirus

286. Single-stranded linear DNA genome, naked icosahedral symmetry

287. Single-stranded linear RNA genome with negative polarity, helical symmetry, enveloped

288. Single-stranded linear RNA genome with negative polarity and eight unique segments, helical symmetry, enveloped

289. Double-stranded linear RNA genome, ten segments, naked icosahedral symmetry

290. Single-stranded linear RNA genome with negative polarity, helical symmetry, enveloped and "bullet-shaped"

291. Double-stranded linear DNA genome, naked icosahedral symmetry

ANSWERS AND TUTORIAL ON ITEMS 286-291

The answers are: **286-D; 287-J; 288-I; 289-F; 290-M; 291-O**. The genome of the parvoviruses (D) may exist as either positive or negative polarity within the capsid. Replication generates large numbers of both molecules.

The reoviruses (F) have ten segmented double-stranded RNA as genome. Each strand encodes one or more unique viral antigens. The viruses also have a unique double capsid structure which is uncoated within the phagosomes of infected cells.

Orthomyxovirus (I) have eight single-stranded RNA segments of negative polarity. Each encodes a unique viral antigen. Genetic assortment, termed genetic shift, of the hemagglutinin-encoding segments between influenza viruses of similar or different hosts may result in the emergence of modified viruses with new virulence properties. Smaller point mutations, termed genetic drift, may result in less profound antigenic changes in the virus.

Paramyxoviruses (J) cause several of the common childhood illnesses such as measles and mumps. They are enveloped viruses with potent hemagglutinin antigen and fusion proteins which take part in the infectious process. The symmetry of these viruses is helical even though their overall structure in spherical.

Rhabdoviruses (M) have a characteristic bullet-shape. They are helical viruses with a single, negative-stranded genomic RNA. The envelope hemagglutinin appears to bind with the acetylcholine receptor antigen on neuronal cells.

Adenoviruses (O) are double-stranded DNA viruses which encode transforming antigens. There are over 100 serotypes of these viruses. A spike polypeptide complex extends from each of the twelve pentameric centers of the naked icosahedron structure. Certain host cells, such as hamster cells, are transformed at a higher frequency than mouse cells by different adenoviruses. Most human infections result in ocular and respiratory infections.

Items 292-296

Several conventional viruses and unconventional agents cause slow dcgenerative diseases of the central nervous system. Match the following viruses or agents with the disease listed.

(A) Measles virus
(B) Rubella virus
(C) Human immunodeficiency virus
(D) Polyoma virus
(E) Prions

292. A person is diagnosed with a progressive multifocal leukoencephalopathy. A personal history indicated childhood exposure to a togavirus.

293. A 17 year-old girl is diagnosed with subacute sclerosing panencephalitis. Her CSF has a high titer against a paramyxovirus which can be rescued by fusion with permissive cells. Viral particles are absent from a brain biopsy specimen.

294. An 18 year-old girl has a progressive demyelinating disease. Her serum and CSF show high titer against JC virus.

295. A 25 year-old man suffers from bacterial and fungal infections. Purple skin lesions cover his legs. He also has loss of memory and he exhibits wide mood changes.

296. Creutzfeldt-Jakob disease

ANSWERS AND TUTORIAL ON ITEMS 292-296

The answers are: **292-B; 293-A; 294-D; 295-C; 296-E**. Measles virus (A) infection in infants, also called rubeola, usually results in a relatively mild form of the disease. Complications associated with bacterial superinfections can include otitis media, pneumonia, mastoiditis or sinusitis. The viremia associated with the infection may also involve the central nervous system resulting in a long-term fatal encephalitis called subacute sclerosing panencephalitis (SSPE). Research has shown that the virus develops a mutation in the matrix protein resulting in a form of latency. A cell-mediated immune response contributes to the destruction of tissue in the CNS leading to neurological deterioration of both gray and white matter in the brain. Most patents affected with SSPE are ages 2-10 years. Vaccination against measles can reduce the incidence of SSPE, however there is no treatment for the disease.

Rubella virus (B), also called the German measles virus, may produce congenital complications which may result in neurological damages later in life called progressive multifocal leukoencephalopathy (PML). These congenital complications often result from infections of the pregnant mother if she is infected with the virus during the first trimester of pregnancy. There is no treatment for PML.

Human immunodeficiency virus (C) (HIV) infection affects the central nervous system of some acquired immunodeficiency syndrome (AIDS) patients. CNS involvement manifestations of viral infection may range from an inapparent infection to marked changes in mood and memory lapses. All of this may occurs in addition to various opportunistic infections and Kaposi's sarcoma, a skin cancer.

The human polyoma viruses (D) include JC and BK viruses. JC virus in humans can cause a demyelinating disease, also sometimes referred to as a progressive multifocal leukoencephalopathy. The disease develops frequently in immunodeficient individuals. JC appears to grow preferentially in the oligodendrocytes of the CNS.

Prions (E) are currently viewed as an infectious self-replicating proteins which have been altered. The altered state of the prion protein (PrP) results in a slow developing degenerative disease, such as Creutzfeldt-Jacob disease and Gerstmann-Straüssler disease in humans. The incubation period may exceed ten years. Kuru (in humans) and scrapie (in cattle and sheep) are also caused by the same prions. Prion-caused degenerative diseases produce a fatal spongiform encephalopathy. Neither humoral or cellular immune responses are activated by prions.

Items 297-302

The detection of some viral infections is facilitated by using culture cells which may produce a cytopathic effects (CPE) characteristic of the unknown virus. Match the following viruses in the answers with the listed disease description and cytopathic effects in the items below.

(A) Herpes simplex virus
(B) Smallpox virus
(C) Reovirus
(D) Measles virus
(E) Rabies virus
(F) Poliovirus
(G) Retrovirus

297. A 36 year-old auto mechanic has several small lesions on the foreskin of his penis. Fluid taken from one lesion was placed on primary rabbit kidney cells which after three days showed giant cell formations with nuclear inclusions.

298. A U.S. Army nurse, on duty in an isolated village in Somalia, discovered an old man with large vesicular lesions on his arms and legs. His palms and feet also showed the lesions. Fluid was taken from several of the uniform crop of lesions and shipped to a lab. The specimen produced lysis and cytoplasmic inclusions in cultures mouse fibroblast cells.

299. A 9 year-old boy developed a fever and faint red rash on his face and chest. Small white blotches surrounded by reddened tissue lined the insides of his cheeks. A scraping of the oral lesions was placed on human cell cultures. After 3 days, cell fusion was evident along with both cytoplasmic and nuclear inclusions. The infected cells also adsorbed erythrocytes placed in the flask.

300. A 27 year-old woman was diagnosed with hairy cell leukemia. Her leucocytes were collected and used to infect hamster cells in culture. Some of the infected culture cells formed foci, but the efficiency was very poor.

301. A 45 year-old postman was bitten by an irate poodle. The owner of the pet had noted unusual aggressiveness in her dog after a recent trip to visit relatives on a farm. The bite became inflamed. A biopsy was stained and examined. It showed cytoplasmic inclusions, similar to negri bodies. The specimen did not cause CPE on cultured mouse cells.

302. A 10 year-old boy traveling with his parents in Jamaica had experienced a viral gastroenteritis. He later developed numbness in his left arm. Fecal specimen showed no parasites. Samples filtered to remove all bacteria caused plaques of cell lysis on cultures human and mouse cells.

ANSWERS AND TUTORIAL ON ITEMS 297-302

The answers are: **297-A; 298-B; 299-D; 300-G; 301-E; 302-F**. Herpes simplex virus type 1 and 2 (A) will produce genital infections. The lesions are indistinguishable. Infected cell cultures produce fused cells depending on the properties of the isolate. Infected cells also show nuclear inclusions of assembling virions.

The last reported case of naturally-occurring smallpox virus (B) took place in Somalia in 1977. The infected man, named Ali Mao Moallin, still lives (1993). Though highly unlikely, this disease may still exist in desolate areas of that country. Humans are the natural reservoir for the variola (smallpox) virus. It produces cytoplasmic inclusions in a variety of culture cells.

Measles virus (D) causes a characteristic petechial rash and lesions on the inner cheek called Koplik's spots. Exfoliated cell scrapings from the lesions contain active virus. Inoculated into cell cultures, measles virus produced both cytoplasmic and nuclear inclusion.

Rabies virus (E) is difficult to isolate from the initial wound. It infects muscle cells at the site of the bite and then enters the nerve endings through neuromuscular junctions. Infected cells develop cytoplasmic inclusions, called Negri bodies, of viral assembly.

The polioviruses (F) are enteroviruses. Large quantities are shed in the feces. Membrane sterilized fecal samples placed on culture cells would produce plaques of cellular lysis. The viruses assemble in the cytoplasm of infected cells prior to lysis.

The human retrovirus (G), called the human T lymphotropic virus II, is associated with hairy cell leukemia. The virus, along with a related human retrovirus, HTLV-II, caused adult T cell leukemia. The viruses grow poorly in cell cultures. Infected transformed cells become immortal and form foci which are rapidly growing cells which do not exhibit contact inhibition.

CPE in infected culture cells may be observed by simple microscopic stains; however, in order to confirm the presence of a particular viral antigen, a specific antibody marker is often used.

Items 303-306

For several weeks, children in a local elementary school had been developing bright red body rashes which became lacelike or reticular over a period of 1-2 weeks accompanied by bright blotches or streaks on the face. In some cases, the reddening resembled a slapped cheek. Fever was absent or low and the students did not seem ill.

303. The most likely diagnosis is

(A) roseola
(B) fifth disease
(C) rubella
(D) rubeola
(E) infectious mononucleosis

304. The infectious agent causing the illness is believed to be a

(A) mycoplasmal organism
(B) virus
(C) fungus
(D) parasitic worm
(E) gram-negative rod

305. Transmission of the agent among the young school children most likely occurs by

(A) respiratory droplet
(B) head lice
(C) ingesting contaminated water or foods
(D) contaminated environmental surfaces
(E) infected pets

306. Laboratory diagnosis is confirmed by the use of

(A) differential agars
(B) selective host cell lines and CPE development
(C) DNA hybridization using specific probes
(D) microscopic examination of biopsy material
(E) examination of blood smears

ANSWERS AND TUTORIAL ON ITEMS 303-306

The answers are: **303-B; 304-B; 305-A; 306-C**. Fifth disease, also called erythema infectiosum or academy rash, is believed to be caused by parvovirus B19. The infection, occurring frequently among young children, produces a mind illness lasting 1-2 weeks, but which may recur weeks later over a longer period. The infection in pregnant women may result in profound complications such as stillbirths or anemia in the fetus.

Diagnosis is based on symptoms, however several other viral infections of children, such as roseola and measles, may produce similar rash patterns. However the presence of DNA sequences from the causative agent, parvovirus B19, is characteristic of fifth disease.

CHAPTER III

IMMUNOLOGY

Items 307-311

The mother of an 8 month-old son brought him to the pediatrician for the second time in a month. The child appeared to have developed a strep throat infection somewhat similar to one he previously experienced. The child experienced a high fever and extreme irritability due to his condition. Bacterial cultures grew β-hemolytic Streptococcus pyogenes. The total level of IgG, IgA and IgM in his serum was only 65 mg/dL. Antibiotics were prescribed to arrest the bacterial infection.

307. The recurrent pyogenic infections experienced by the boy, along with his low overall serum immunoglobulin level suggests

    (A) persistent infections by drug-resistant bacteria
    (B) an immunodeficiency
    (C) the mother is a carrier of the bacteria
    (D) misdiagnosis of earlier infections
    (E) age-related immunodeficiency

308. A disease which results in the markedly reduced levels of all immunoglobulins is

    (A) Bruton's agammaglobulinemia
    (B) DiGeorge syndrome
    (C) Wiskott-Aldrich syndrome
    (D) ataxia-telangiectasia
    (E) Chédiak-Higashi syndrome

309. The defective gene causing this disease is located on which chromosome?

    (A) Chromosome number 13
    (B) Chromosome number 18
    (C) Chromosome number 20
    (D) The X chromosome
    (E) The Y chromosome

310. The basis of the low total immunoglobulin level is a

(A) T cell deficiency
(B) B cell deficiency
(C) complement deficiency
(D) phagocyte deficiency
(E) disease acquired deficiency

311. Treatment of the patient, in addition to antibiotics, would include

(A) passive immunization
(B) active immunization
(C) transfusions
(D) immunosupressive drugs
(E) cytokine administration

ANSWERS AND TUTORIAL ON ITEMS 307-311

The answers are: **307-B; 308-A; 309-D; 310-B; 311-A**. Recurrent pyogenic bacteria infections in infants from organisms such as Streptococcus may indicate an underlying immunodeficiency. Bruton's agammaglobulinemia, also called X-linked agammaglobulinemia, results from a B cell deficiency. B cell deficiency results in a reduction in B cell function and concomitant reduction in immunoglobulins, secreted by plasma cells which are derived from B cells. It is X-linked and therefore expressed only in males (XY). The serum of infants show a uniform reduction in all immunoglobulin levels. Infections, starting around 6 months of age, occur with increasing frequency as maternal IgGs, transferred across the placenta during the fetal period, are depleted from the infants circulatory system after birth.

In a related disorder, X-linked hyperimmunoglobulin M syndrome (HIM), patients produce normal levels of IgM. B cells fail to "switch" into producing other immunoglobulins. HIM is believed to be due to a defective CD40 ligand on T cells. CD4 ligand is essential for activated helper T cell interaction with B cells.

One treatment used in cases of agammaglobulinemia is pooled gamma globulin. This form of passive immunization provides the infant with higher levels of mixed antibody, some of which may cross react with a variety of commonly occurring pathogens, resulting in a reduction of infections. The genetic nature of the disease currently precludes any permanent cure. Future advances in gene therapy may one day provide alternative approaches.

Items 312-315

The mother of an 8 month-old son noted an unusual recurring illnesses consisting of upper respiratory tract infections, most likely of viral origin. No members of the household had similar infections, however some visiting relatives showed signs of infections. The infant had consistently congested nasal passages and wheezed with each breath. A physical examination detected mild facial hypoplasia. Microbiological studies did not indicate any bacterial infections. Serum immunoglobulin levels were within the average range. Blood chemistry indicated hypocalcemia.

312. A disease which follows this particular pattern of immunodeficiency in infants is

(A) Bruton's agammaglobulinemia
(B) DiGeorge syndrome
(C) Wiskott-Aldrich syndrome
(D) ataxia-telangiectasia
(E) Chédiak-Higashi syndrome

313. The immune dysfunction is due to

(A) congenital viral infection
(B) toxin effect during pregnancy
(C) bone marrow malformation
(D) thymus and parathyroids malformed
(E) defect in stem cell differentiation

314. An examination of the infant's immune functions would most likely indicate a deficiency of

(A) B cell number and function
(B) T cell number and function
(C) macrophage number and function
(D) plasma cell number and function
(E) complement titer and function

315. Which of the following would be most effective in overcoming the deficiency?

(A) blood transfusion
(B) cytokine therapy
(C) thymic transplant
(D) bone marrow transplant
(E) gamma globulin treatment

ANSWERS AND TUTORIAL ON ITEMS 312-315

The answers are: **312-B; 313-D; 314-B; 315-C**. This child suffers from DiGeorge syndrome. This immunodeficiency disease is caused by congenital hypoplasia of IIIrd and IVth pharyngeal arch derivatives. Hypoplasia of the face, the thymus and the parathyroid glands are all associated with DiGeorge syndrome.

Patients with DiGeorge syndrome often have near normal levels of immunoglobulins for their age but significant decreases in T cell numbers and a concomitant relative increase in the percentage of B cells. The expression of this disease is highly variable ranging from mild to profound. Thymic hypoplasia or complete lack of a thymus results in a decrease in T cell numbers and a reduction in the cell-mediated immune response. In some instances, thymus-dependent areas of the lymph nodes and spleen also show variable T cell depletion. The hypocalcemia is caused by depressed secretion of parathormone by the hypoplastic or absent parathyroid glands

The eradication of protozoal, fungal and viral infections depends on the function of an adequate cell-mediated immune (CMI) response. The essential T cells of the CMI mature in the thymus. Malformation of the thymus and parathyroid results in a major immune dysfunction resulting in negligible T cell populations. Cytotoxic T cells and helper T cells act to kill host cells infected with intracellular pathogens. Both cell populations also produce essential cytokines which stimulate clonal expansion of stimulated (activated) T cells and B Cells.

Transplants of thymic tissue have had some limited success in correcting this deficiency. Claimed benefits from thymic transplants are difficult to evaluate, given the highly variable nature of the initial immunodeficiency state in DiGeorge syndrome. (congenital thymic aplasia)

Items 316-321

(A) Selective immunoglobulin deficiency
(B) Chronic mucocutaneous candidiasis
(C) Severe combined immunodeficiency disease (SCID)
(D) Adenosine deaminase and nucleotide phosphorylase deficiency
(E) Wiskott-Aldrich syndrome
(F) Hereditary angioedema
(G) Chronic granulomatous disease
(H) Job's syndrome
(I) Common variable hypogammaglobulinemia
(K) Acquired immunodeficiency syndrome
(L) Chédiak-Higashi syndrome
(M) Ataxia-telangiectasia
(N) Thymic aplasia
(O) X-linked agammaglobulinemia

Match the description of immunodeficiencies in the items below with the most appropriate disease above.

316. Inability to mount a IgM response to bacterial capsule polysaccharides

317. Inability to produce a particular immunoglobulin, due to recurrent bacterial infections during teen and early adult years

318. High levels of IgE, but inability of neutrophils to respond to chemotaxis stimuli

319. Overall cell-mediated immune dysfunction resulting from CD4+ T cell depletion

320. Deficiency in C1 esterase inhibitor resulting in the indirect generation of vasoactive kinins

321. Neutrophils fail to respond to chemotactic signals and therefore do not respond to the presence of pathogens

ANSWERS AND TUTORIAL ON ITEMS 316-321

The answers are: **316-E; 317-I; 318-H; 319-K; 320-F; 321-C**. Immunodeficiencies are either congenital or acquired. Each type may adversely alter the key elements of the immune system (T cells, B cells, antigen processing cells and cytokines). The imbalance caused by the deficiencies will have profound effects on the ability of the body to survive the assault posed by different pathogens and tumors.

Acquired immunodeficiencies include: common variable hypogammaglobulinemia (I), in which B cells do not produce adequate levels of IgG; and acquired immunodeficiency syndrome (K), caused by the viral destruction of CD4+ helper T cells.

Congenital immunodeficiencies can affect:

1. B cells, as in X-linked agammaglobulinemia (O), where B cells fail to mature due to a tyrosine kinase deficiency; or, selective immunoglobulin deficiency (A), where B cells fail to switch in the production of IgA, but IgG and IgM levels are normal.

2. T cells, as in thymic hypolasia (N) (DiGeorge syndrome), where the thymus fails to develop; or, chronic mucocutaneous candidiasis (B), where there is a specific T cell failure to specifically recognize the antigens of Candida albicans.

3. Combined B cells and T cells, as in severe combined immunodeficiency disease (C), where an individual fails to inherit the class II MHC antigens; adenosine deaminase and nucleotide phosphorylase deficiency (D), where the absence of certain enzymes leads to the accumulation of adenosine and adenosine metabolites, leading eventually to lymphocyte toxicity; Wiskott-Aldrich syndrome (E), where IgM is not produced; and, finally, ataxia-telangiectasia (M), where lymphopenia and IgA deficiency occurs as early as age 2.

4. Antigen processing cells, as in chronic granulomatous disease (G), where neutrophils lack NADH oxidase (needed to kill pathogens); Chédiak-Higashi syndrome (L), where the lysosomes of neutrophils fail to release their contents into phagosomes; or, Job's syndrome (H), where neutrophils do not respond to chemotactic signals.

5. Cytokine deficiencies, as in hereditary angioedema (F), where the complement component enzyme C1 esterase inhibitor in not produced, leading to the excess production of vasoactive kinins; or, C2/C4 deficiencies, where an autoimmune disease results.

Each immunodeficiency syndrome is associated with the occurrence of infections by similar groups of pathogens. The pattern of infection and studies of immune function are used in the diagnosis of the illness. Treatment of the different immunodeficiencies vary with the syndrome, age and the patient's responsiveness to drugs, immune modulators or tissue transplants.

Many viral infections (other than HIV and related viruses) cause transient immunosuppression due to viral replication in different B cell or T cell subpopulations. Examples of this include Epstein-Barr and measles viruses.

Items 322-326

A passively sensitized human volunteer, when challenged with allergen intradermally, displayed a typical skin reaction due to an immediate hypersensitivity (Type I) reaction. If the injection of sensitizing antigens were preceded by the injection (at the same site) of Fc fragments derived from human IgE molecules, no reaction would occur upon challenge with the allergen.

322. The cells in the human volunteer which were sensitized by the passively transferred antibodies were

(A) T cells
(B) B cells
(C) mast cells or basophils
(D) neutrophils
(E) platelets

323. Eosinophil chemotactic factor (ECF-A), released during the development of the skin reaction, would attract predominantly which of the following to the reaction site?

(A) macrophages
(B) T cells
(C) granulocytes with large red granules
(D) granulocytes with minute blue-grey granules
(E) granulocytes with prominent metachromatic granules

324. ECF-A is tetrapeptide found preformed within

(A) macrophages
(B) T cells
(C) eosinophils
(D) neutrophils
(E) mast cells

325. The role of the IgE Fc fragments in blocking the skin reaction can be explained by the fact that they

(A) interact with the allergen and prevent it from gaining access to the sensitized cells
(B) interact with the IgE, rendering their antigen binding sites unavailable for binding antigen
(C) interact with receptors on the surface of cells and block the attachment of IgE
(D) activate complement
(E) lyse sensitized cells.

326. One of the primary mediators released in the reaction is

(A) migration inhibitory factor (MIF)
(B) interferon
(C) cathepsin
(D) histamine
(E) interleukin

ANSWERS AND TUTORIAL ON ITEMS 322-326

The answers are: **322-C; 323-C; 324-E; 325-C; 326-D**. Individuals react differently to exposure to certain antigens. Antigens that give rise to hypersensitivity (also called allergy) are referred to as allergens. The body may react at least four different ways or types (e.g. Types I, II, III and IV) to exposure to allergens. The immediate Type I, or anaphylactic, reaction is demonstrated in the case presented.

In the Type I hypersensitivity reaction, the primary mediator is antibody, namely IgE. The initial exposure to an allergen causes the production of IgEs which are selectively bound by their Fc portions to mast cells and basophils. This sensitizes the cells to that particular allergen. Upon a second exposure to the allergen, sensitized cells, with exposed Fab portions, bind the allergen. This causes immediate cross-linkages between bound antibodies and degranulation within the mast cells. Degranulation is the release of preformed granular materials from the mast cells. The materials released consist of certain mediators of anaphylaxis.

Histamine is a primary anaphylaxis mediator. It causes vasodilation, increased capillary permeability and smooth muscle contraction. Another mediator is eosinophil chemotactic factor, a tetrapeptide, which once released attracts eosinophils, which in turn contribute to the overall reaction.

Other mediators of anaphylaxis includes serotonin, slow-reacting substance of anaphylaxis, prostaglandins and thromboxanes. Other cell types, such as platelets, carry these mediators.

Items 327-333

(A) IgA
(B) IgD
(C) IgE
(D) IgG
(E) IgM
(F) J chain protein
(G) Secretory component

Match the following descriptions of properties or functional roles in the items below with the most appropriate molecule listed in the answers above.

327. Function presently unknown, found on the surface of some B cells in cord blood, may involve memory cell function

328. Mediates the immediate hypersensitivity response

329. Early response immunoglobulin, most efficient immunoglobulin, sometimes joined by J chain

330. Predominant immunoglobulin, long-lasting, can cross the placenta

331. Secretory or mucosal immunoglobulin, joined by J chain

332. In addition to IgG, it also fixes complement

333. Added to immunoglobulins secreted by epithelial cells, it covers the hinge region of immunoglobulins, rendering them less susceptible to enteric proteolysis

ANSWERS AND TUTORIAL ON ITEMS 327-333

The answers are: **327-B; 328-C; 329-E; 330-D; 331-A; 332-E; 333-G**. B cells secrete four major classes of immunoglobulins. Each serves critical functions in protecting the host from the pathogens and tumors. However, their beneficial roles are sometimes subverted by pathogens, tumors and even the host itself.

The immunoglobulins differ in their molecular composition and function. IgG (D) inactivates bacteria, viruses and other parasites by coating (opsonizing) them and thus rendering them easily engulfed and degraded by phagocytic cells. IgG is the most abundant immunoglobulin. It is maintained in the blood for extended periods and fixes complement. It protect the fetus by its ability to cross the placenta.

IgM (E) can form pentameric complexes with the J (joining) chain protein (F). It is the first immunoglobulin expressed after exposure to an antigen. It has the highest binding affinity of all immunoglobulins.

IgA (A) exists as monomers or dimers in the serum or as secreted dimers in tears, milk, mucus and secretions of other mucosal surfaces. Dimerization involves an association with the J chain protein. The secretory component (G) is an IgA receptor on the surface of many epithelial cells (e.g. mammary gland acinar cells). It binds IgAs (synthesized by plasma cells in the connective tissue domains underlying the epithelial layers) covering the hinge region (proteolysis site). The secretory component is added to IgA dimers as they pass through epithelial cells and enter secretions. Complete secretory IgAs are thus less susceptible to enteric proteolysis. This is part of the reason why breast fed infants are less susceptible to gastrointestinal diseases.

IgE (C) is involved with immediate hypersensitivity or allergic responses expressed by the release of mediators from mast cells and basophils.

The function of IgD (B) remains unresolved. It is found on the surface of B cells and may serve in the role of an antigen receptor. The presence of IgD has also been associated with memory cell function.

Items 334-336

A patient entered the hospital suffering from an inner ear infection. A blood culture showed the presence of Staphylococcus aureus. The patient's cell count appeared normal except for a CD4+ count of 150 cells/mm^3. A second reading showed 275 cells/mm^3 (normal CD4+ range 600-1200 cells/mm^3).

334. The low CD4+ T cell count

(A) may indicate an underlying immunosuppression
(B) is normal for individuals with Staphylococcus infections
(C) is within the error range of the assay
(D) fluctuates widely in all healthy individuals
(E) indicates AIDS

335. CD4+ cells perform all of the following functions **EXCEPT**:

(A) stimulate B cells to becoming antibody-producing cells
(B) activate CD8+ T cells
(C) assist in delayed type hypersensitivity
(D) secrete lymphokines
(E) lyse tumor cells

336. CD8+ cells perform all of the following functions **EXCEPT**:

(A) respond to CD4+ cell signals
(B) lyse virus-infected cells
(C) assist in delayed type hypersensitivity
(D) secrete lymphokines
(E) lyse tumor cells

ANSWERS AND TUTORIAL ON ITEMS 334-336

The answers are: **334-A; 335-E; 336-C**. A bacteremia is not normally associated with T cell immunodeficiency, as opposed to transient B cell deficiency. The low CD4+ helper T cell count is significantly below the normal range. The assay has inherent errors and therefore repeat measurements are routinely performed before drawing diagnostic conclusions. CD4+ cells serve many critical functions in both the stimulation of B cells to form plasma cells and activation of CD8+ cells. This is accomplished, in part, through the secretion of different cytokines (interleukins) and other factors. Helper cells also respond to signals emitted by macrophages and possibly other T cells. They help in B cell differentiation into plasma cells, participate in delayed type hypersensitivity reactions and stimulate cytotoxic T cells. CD8+ cells lyse virus-infected cells, lyse tumor cells and secrete lymphokines.

The presence of the CD4+ marker on T cells functions in conjunction with a competent set of other T cell markers such at the CD3, CD40 and α/β T cell receptors. Recent evidence suggests that T cells may loose functionality due to factors secreted by animal viruses, tumors and possibly other parasites. In the future, functional T cell assays will become more important in evaluating immunodeficiency.

<u>Items 337-339</u>

Certain individuals react violently to antigens that elicit little response in other individuals. Bites by the common gnat may cause an immediate and profound reaction. An old treatment used to reduce or eliminate the reaction used antigens isolated from gnats to desensitize patients. The procedure has drawbacks.

337. Desensitization by injection of gnat antigens was effective because

(A) it produced excess circulating antigen
(B) it saturated all IgE attached to mast cells
(C) it caused excess production of IgE
(D) it caused excess production of IgG
(E) the small complexes formed do not cause degranulation

338. All of the following compounds have been shown effective in countering mediator release **EXCEPT**:

(A) epinephrine
(B) antihistamines
(C) cromolyn sodium
(D) corticosteroids
(E) interferon

339. Sensitivity to the drug penicillin is an example of

(A) Type I hypersensitivity
(B) Type II hypersensitivity
(C) Type III hypersensitivity
(D) Type IV hypersensitivity
(E) Type V hypersensitivity

ANSWERS AND TUTORIAL ON ITEMS 337-339

The answers are: **337-E; 338-E; 339-A**. The effects of immediate (Type I) hypersensitivity may be addressed through desensitization or drug therapy. Desensitization can involve the short-term administration of small amounts of the allergen. The amounts of allergen present cause only minimal IgE complexes to form without massive degranulation. Once IgE levels are lowered, there is reversion to the sensitized state within a week or two. This approach is applicable when the allergen is a necessary therapeutic agent such as in special cases of penicillin therapy. Longer term administration of the allergen introduces sufficient amounts of the allergen in order to elicit an immune (IgG) response. The latter response functions as a blocking antibody designed to remove the allergen prior to interaction with sensitized mast cells.

Compounds which counter the effects of histamines include epinephrine, antihistamines, corticosteroids and cromolyn sodium. The interferons are not effective on differentiated cell populations.

Items 340-345

The serum of most individuals contains 20 proteins which constitute the complement system. By definition these proteins "complement" the actions of the antibody-antigen reaction and also serves other important functions.

(A) C1
(B) C3
(C) C3b
(D) C3BD
(E) C4b,2b
(F) C5
(G) C5a
(H) C5a5b
(I) C567
(J) C5b6789

Match the functional description of a complement component in the items below with the most appropriate complement component/complex in the answers above.

340. Directly involved with the opsonization of bacteria

341. Causes degranulation of mast cells

342. Causes lysis of tumor cells

343. Weakly attracts leucocytes (chemotactic factor)

344. Convergence point of both the classic and alternative pathways of the complement system

345. Initial membrane attachment complex

ANSWERS AND TUTORIAL ON ITEMS 340-345

The answers are: **340-C; 341-G; 342-J; 343-I; 344-F; 345-H**. The complement system consists of an assemblage of 20 proteins synthesized in the liver. The interaction of the different components are affected by the presence of antibody-antigen complexes, component enzymes and competitive binding antigens in the serum. The alternative pathway is used when the new antigens are introduced into the animal and no prior antibodies are present. The classical pathway is activated by antibody-antigen complexes within the system. C5 is the component situated at the convergence of both pathways. It is produced by the respective C5 convertases.

Bacteria, for example, are opsonized by the presence of bound C3b on phagocytic cells. Mast cell degranulation is mediated by several complement components including C3a, C4a and C5a. The result is the release of histamine causing an increase in vascular permeability and smooth muscle contraction. Chemotaxis of leucocytes is a function of C5a and C567, however the C5a is more effective.

An essential point in the effectiveness of complement in the lysis of bacteria or tumor cells is the binding of C5a5b to the membrane and finally lysis after insertion of the C5b6789 complex into cell membranes, rendering them porous.

Items 346-351

A happily married couple enjoyed the friendship of a male whom they had known for many years. Unexpectedly the wife became pregnant. When she told her husband, he curtly informed her that he didn't believe that he was the father of the child. The friend also denied paternity. The HLA-A and HLA-B phenotypes of the adults were:

Husband:	A 1, 3 and B 7, 8
Wife :	A 3, 9 and B 12, 27
Friend :	A 1, 9 and B 8, 12

346. Which of the following phenotypes is likely to correspond to the child whose parents are the husband and wife?

   (A) A 1, 3; B 7, 12
   (B) A 1, 3; B 12, 27
   (C) A 1, 9; B 7, 8
   (D) A 3, 9; B 12
   (E) A 1 ; B 7, 12

347. Which of the following phenotypes is NOT possible for a child fathered by the friend?

   (A) A 1, 3; B 8, 12
   (B) A 9 ; B 12
   (C) A 1, 9; B 8, 27
   (D) A 3, 9; B 12, 27
   (E) A 9 ; B 8

348. The child born to the mother had the phenotype of A 1, 9; B 8, 12. Who was the male parent?

   (A) It was the wife's husband
   (B) It was the friend
   (C) It could have been either the husband or the friend
   (D) It was definitely a third male
   (E) None of the above is possible

349. All of the following properties of class I MHC antigens are true **EXCEPT**:

(A) They are glycoproteins
(B) They are derived from 3 possible loci
(C) They consist of one main protein
(D) They are complexed with β_2-microglobulin
(E) They are present on only certain cells

350. All of the following properties of class II MHC antigens are true **EXCEPT**:

(A) They are glycoproteins
(B) They are derived from 3 possible loci
(C) They are two main proteins
(D) They are complexed with β_2-microglobulin
(E) They are present of all cells

351. A graft from the mother to the child would be a(n)

(A) xenograft
(B) autograft
(C) allograft
(D) syngeneic graft
(E) heterograft

ANSWERS AND TUTORIAL ON ITEMS 346-351

The answers are: **346-A; 347-E; 348-C; 349-E; 350-E; 351-C**. The antigens of the major histocompatibility complex (MHC) determine the survivability of tissue and organ transplants. Typing of the class I MHC can also be used to show relatedness of individuals since the alloantigens are present on all cells of the body and the genes which encode them segregate as simple Mendelian traits. The A, B and C loci of class I MHC are very diverse with 20, 40 and 8 possible variations being possible at each locus, respectively.

The case history illustrates the significance of phenotype expression and haplotype identity in assessing possible relatedness of individuals. Each HLA locus has two possible inherited isotypes located on each chromosome. Offspring inherit a marker from each biological parent in a random manner. At the HLA A locus, the husband can contribute either marker 1 or 3 to the child. The wife can contribute 3 or 9 and the friend can contribute 1 or 9. If the parents contribute a shared marker, e.g. marker 3 between husband and wife, then the child exhibits only the one marker since both of the child's chromosomes have the same marker. Determining possible phenotypes involved reconstructing all of the possible outcomes resulting from a particular paternal/maternal pair.

The closer the class I MHC match between donor and recipient of organ transplants, the greater the survivability of the transplant. Transplants between identical twins (syngeneic grafts) are a perfect match and are fully accepted. Allografts are grafts between different members of the same species. Xenografts (also called heterografts) are grafts between different species e.g. chimpanzee to human organ transplants are xenografts. Autografts are grafts within the same individual, as in coronary artery bypass surgery. Autografts and syngeneic grafts are always fully accepted without graft-versus-host rejection.

Class I and II MHC are composed of different glycoproteins complexed with other cellular proteins. β_2-microglobulin is an important factor in each complex formation. Class I MHC are found on all nucleated cells whereas class II are only found on specialized cells such as macrophages, B cells, dendritic cells and Langerhans cells of the skin. The main proteins of both MHC classes, 1 for class I and 2 for class II, are highly polymorphic with constant and hypervariable regions.

Items 352-355

A married couple decided to have a second child. The obstetrician informed them of possible complications due to differences in their red blood cell antigens. Their first child, a son (blood type A +), was born healthy. The typing results showed the following results

	Blood type
Father	B, Rh +
Mother	A, Rh -

352. The second child would be at possible risk if his/her blood type were

(A) A, Rh -
(B) B, Rh -
(C) AB, Rh -
(D) A, Rh +
(E) O, Rh -

353. The subsequent immunological disease against red cells will affect

(A) all future pregnancies
(B) only male fetuses
(C) only female fetuses
(D) only Rh(D) + fetuses
(E) only Rh(D) - fetuses

354. Sensitization of the mother probably occurred during the

(A) conception of her first child
(B) the second trimester of her first pregnancy
(C) delivery of her first child
(D) first trimester of her first pregnancy
(E) first trimester of her second pregnancy

355. A test used to determine whether the mother has been sensitized against blood antigens is the

(A) complement fixation test
(B) mixed lymphocyte test
(C) PCR
(D) ELISA
(E) Coombs test

ANSWERS AND TUTORIAL ON ITEMS 352-355

The answers are: **352-D; 353-D; 354-C; 355-E**. Red blood cells contain different surface antigens that are important in transfusion and autoimmune reactions. The ABO blood groups differ by the addition of a sugar residue, either galactose and/or N-acetylgalactosamine to a common H antigen. The H antigen consists of a 4-sugar core consisting of N-acetylgalactosamine, glucose, galactose and fucose. The A and B blood groups encode sugar transferases which add the final sugar.

The Rh(D)+ antigen is dominant. Mothers who deliver Rh + infants may become sensitized when blood from the infant crosses the placenta during delivery. All subsequent pregnancies in which the infant is Rh + risk a severe hemolytic disease against the fetus called erythroblastosis fetalis. Maternal antibodies (IgG) against Rh(D)+ antigen can cross the placenta and lyse the red blood cells of the fetus. Sensitized mothers develop an autoimmune response against the Rh(D)+ antigen. A Coombs test is used to detect the presence of such antibodies in the mother's serum. The test uses the mothers serum mixed with washed Rh(D)+, group O, red cells. If agglutination of the red cells occurs with the addition of anti-human IgG, then the mother has circulating anti-Rh(D)+ antibodies.

Items 356-359

A researcher, working on the HIV, set up a double immunodiffusion plate. She isolated the virus from a patient, grew it in human cells and prepared whole-virus antibody in rabbits. The distribution of sera and antigen in the experiment are as indicated:

	Side A	Side B
Center Well:	Patient's serum	Rabbit anti-HIV serum

well # (Side A and Side B)
X=whole HIV
1=human β_2-microglobulin
2=gp-120 protein
3=gp-41 protein
4=p24 protein

Side A Side B

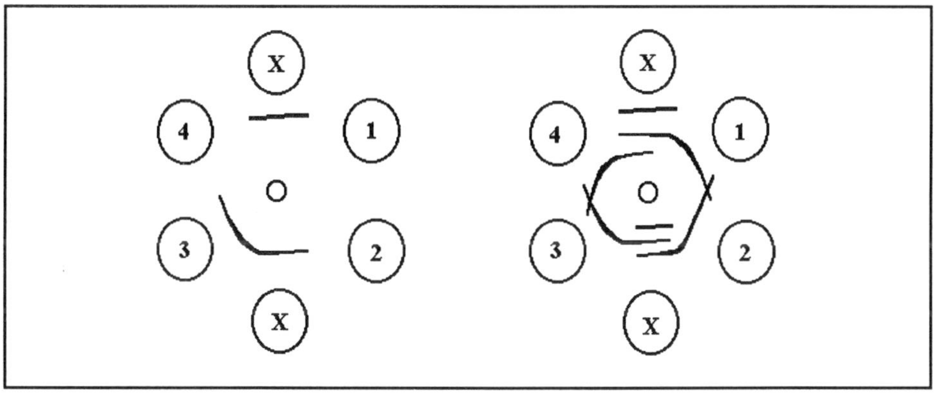

Figure 3.1

356. The pattern on Side A shows

(A) antibody to only one HIV antigen
(B) partial identity between β_2-microglobulin and an HIV antigen
(C) identity between gp-41 and gp-120
(D) identity between gp-41 and β_2-microglobulin
(E) partial identity between gp-41 and an HIV antigen

357. The pattern on Side B shows

(A) antibody to only one HIV antigen
(B) partial identity between β_2-microglobulin and an HIV antigen
(C) identity between gp-41 and gp-120
(D) identity between gp-41 and β_2-microglobulin
(E) identity between gp-120 and β_2-microglobulin

358. A significant difference in Side B is the presence of

(A) anti-gp41 antibody
(b) anti-gp120 antibody
(C) anti-HIV antibody
(D) anti-β_2-microglobulin antibody
(E) anti-p24 antibody

359. The results suggest

(A) HIV codes for β_2-microglobulin
(B) there is a cross reactive antigen in the preparation
(C) human β_2-microglobulin becomes part of the virion composition
(D) rabbit antibody is more sensitive than human antibody
(E) a contaminant in the virus preparation

ANSWERS AND TUTORIAL ON ITEMS 356-359

The answers are: **356-E; 357-B; 358-D; 359-C**. The double immunodiffusion, or Ouchterlony, reactions are a useful laboratory tool for determining qualitative relatedness between antigens or specificities of antibodies. The precipitation reactions formed at the interface of antibody-antigen reactions will indicate nonidentity, partial identity and full identity between reagents, as represented by continuous, single spur or double spur patterns respectively.

The reactions on Side A show partial identity between gp-41 and the whole HIV antigen. This indicates there are additional antigens in the HIV sample not present in the pure gp-41 preparation, as detected by the patient's serum. The immunized rabbit serum detects the viral antigens gp-41, gp-120 and p24 (core protein), as expected. However, there is a significant difference in the reaction of the rabbit serum against the β_2-microglobulin and its partial identity with whole virus, as detected by the rabbit serum. This result is illustrates recent findings that HIV incorporates a large amount of human β_2-microglobulin into its envelope as they mature. Since β_2-microglobulin is "self", the human serum has no antibody against it. Rabbits immunized against the virus easily detect the new "foreign" antigen.

Items 360-366

The surfaces of B cells and T cells possess functionally important receptors (antigens) which determine their immunological roles. Match the following antigens in the answers with the most appropriate description of the structure or function of the antigen in the items below.

(A) T cell receptor
(B) CD3 antigens
(C) CD4 antigens
(D) CD8 antigens
(E) LFA-3
(F) β_2-microglobulin
(G) Class I MHC
(H) Class II MHC
(I) Class III MHC
(J) CD2 antigen
(K) B7
(L) CD28 antigen
(M) CD40 antigen

360. Monomeric heavy chain, MW = 44,000, membrane anchored, three cytoplasmic domains, present on virtually all cells, associated with smaller protein, presents foreign antigen to cytotoxic T lymphocytes

361. Polypeptide of MW = 12,000, associated non-covalently with both class I and II MHC

362. Two distinct polypeptides (α and ß), having variable, diverse and constant segments, found complexed with CD3 proteins

363. Two distinct polypeptides, MW = 34,000 and 29,000, both membrane anchored, two cytoplasmic domains each, presents on B cells, activated T cells and macrophages, restricts helper functions

364. An assemblage of at least four distinct membrane-bound polypeptides, some forming heterodimers, all found in conjunction with the T cell receptor.

365. Two distinct polypeptides, α and ß, MW = 40,000 and 50,000, present on helper and cytotoxic T cells, recognizes MHC embedded antigen

366. Glycoprotein, MW = 50,000, found on mature T cells, natural killer cells and thymocytes, binds to the lymphocyte function antigen-3 (LFA-3)

ANSWERS AND TUTORIAL ON ITEMS 360-366

The answers are: **360-G; 361-F; 362-A; 363-H; 364-B; 365-A; 366-J**. The antigens present on T cells vary in their molecular composition and function. Many possess multiple domains which determine how they will be situated within the cell membrane and interact with other T cell antigens.

The $\alpha\beta$ T cell receptor (A) recognizes the processed antigen fragment presented by the antigen presenting cells through their class II MHC. The T cell receptor interacts with cell surface antigens which functionally define cytotoxic, CD8 (D), or helper, CD4 (C), functions. Class II MHC (H) antigens, present on immunocompetent cells such as B cells, macrophages and activated T cells, function to restrict helper functions. Class I (G) MHC antigens, present on all cells, restrict the targeted killing effects of cytotoxic cells. Both MHC classes form noncovalent associations with the β_2-microglobulin (F) found on T cell surfaces.

The CD3 antigens (B) comprise at least four different membrane bound polypeptides which associate with the T cell receptor in different orientations. The proteins, designated γ, δ, ϵ and ζ , interact individually, as heterodimers and homodimers under different situations. For example, the CD3 complex is thought to be involved in transmembrane signaling. Cell activation via the CD3 complex results in elevation of intracellular Ca^{2+} and phosphorylation of CD3 subunits. The rearrangement of these antigens may be functionally altered in tumor cells rendering them unrecognizable by cytotoxic T cells. The MHC III antigens (I) consist of C2, C4 and B complement components.

Other cell-to-cell interactions must also occur for T cell activation to occur. Interactions such at the binding of LFA-3 (E) (on the surface of T cells) to the CD2 antigen (J) (on the antigen presenting cell) must occur. Blockage of LFA-3 with antibody prevents T cell signaling. The specific ligand for other T cell antigens, such as CD45, has not been identified.

Recent results suggest that antigens found on tumor cells, e.g. the B7 antigen (K), present on the surface of certain mouse tumors, act as a costimulator of cytotoxic T cell activity by binding to the CD28 antigen (L) present on cytotoxic T cells. The costimulation, with MHC I and T cell receptor comprising the primary recognition interaction, induces multiplication of T cells able to recognize the previously ignored target cells.

The CD40 antigen (M), present on B cells, interacts with its corresponding ligand which is present on activated T cells. This allows B cells to switch from expressing IgM to other immunoglobulins. Different cytokines determine which immunoglobulin will eventually be expressed by the activated B cells.

Items 367-371

Various cytokines serve crucial roles in the maturation and stimulation of different cellular subpopulations. Match the cytokine in the answers with the most appropriate description of its primary function in the items below.

(A) IL-1
(B) IL-2
(C) IL-3
(D) IL-4
(E) IL-5
(F) IL-6
(G) IL-7
(H) IL-8
(I) IL-9
(J) IL-10
(K) Gamma interferon (INF-γ)
(L) Transforming growth factor-β

367. Produced by helper T cells, stimulates B cell differentiation and proliferation, also promotes IgA switching

368. Produced by helper T cells, enhances both class I and class II MHC expression, also promotes immunoglobulin switching to IgG and IgE, in mice

369. Produced by a many cell types such as B cells, macrophages and monocytes, enhances natural killer cell activation and stimulates helper T cell activation

370. Produced by macrophages, chemoattractant for neutrophils

371. Produced by platelets, lymphocytes and macrophages, attracts other macrophages and monocytes, induces IL-1 production

ANSWERS AND TUTORIAL ON ITEMS 367-371

The answers are: **367-E; 368-D; 369-A; 370-H; 371-L**. The efficiency of the immune response mechanism is dependent upon direct cell-to-cell contact and indirect chemical signaling by cytokines. A large array of cytokines are produced by leucocytes. They are called interleukins (IL); and, along with gamma interferon (IFN-γ), are the major soluble mediators of immunological function. An evaluation of the interleukins demonstrates multiple regulatory functions. Interleukins are designated by IL- followed by a number. All are proteins and many are glycosylated. They have molecular weights ranging from 15,000 to 60,000. Most have mitogenic properties and each is antigenically unique. Functionally different interleukins share similar roles in stimulating certain B cell or T cell populations.

IL-1 (A) is produced by many cells and has both T cell and B cell activating functions. It also induces IL-2 receptor expression. IL-1 stimulates the cytocidal actions of cytotoxic T cells, macrophages and natural killer cells. IL-2 (B) upregulates lymphokine production and cytotoxic activity of different leucocytes. IL-3 (C) helps to maintain the growth of mast cells and differentiation of monocytes. IL-4 (D) enhances both class I and II MHC expression, stimulates IgG4 and IgE production and stimulates IgE receptor expression on B cells. IL-5 (E) stimulates B cell growth, induces differentiation of eosinophils, stimulates IgA secretion and stimulates B cell growth. IL-6 (F) stimulates IgG secretion, induces the growth of plasma cells and activates B cells. IL-7 (G) stimulates B cell differentiation and increases IL-2 receptor expression. IL-8 (H), produced by macrophages, is a neutrophil chemoattractant. IL-9 (I) stimulates helper T cells. IL-10 (J) suppresses production of certain other lymphokines produced by helper T cells. Transforming growth factor-β (L) is produced by macrophages, lymphocytes and platelets. It acts to attract other macrophages and monocytes and stimulates IL-1 production. It may also down-regulate the inflammatory response. It is important to realize that cytokines usually have multiple roles which may change under different immunological conditions. Many cytokines also act synergistically.

Items 372-376

A young child was immunized with the Salk polio vaccine. A month later, he was given a booster shot of the same vaccine. Studies designed to monitor the child's seroconversion, by measuring both IgM and IgG titers against the vaccine antigen, revealed the serum titer profile shown in Figure 3.2 below:

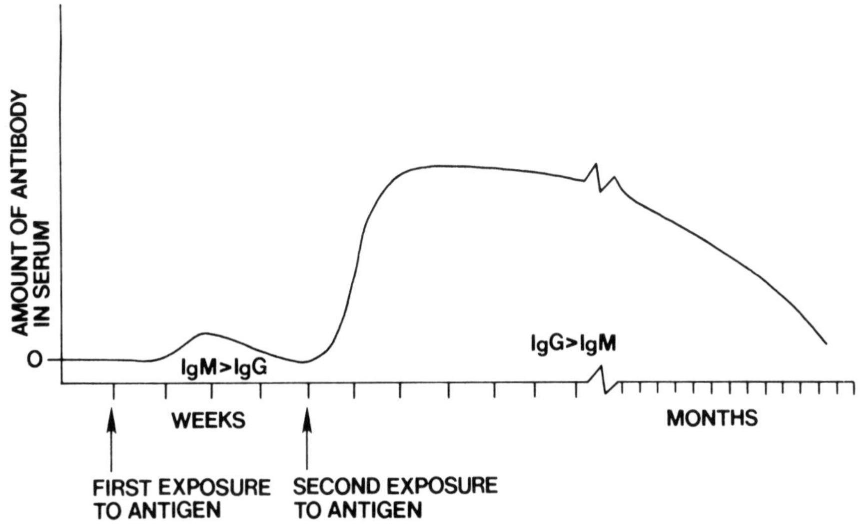

Figure 3.2

372. The heightened level of antibody production following the booster vaccination is commonly referred to as

(A) differentiation
(B) stimulation
(C) memory
(D) recognition
(E) activation

373. The immunoglobulin first produced after first exposure to a new antigen is

(A) IgA
(B) IgM
(C) IgE
(D) IgG
(E) IgD

374. The longest lived immunoglobulin normally found after the booster vaccination is

(A) IgA
(B) IgM
(C) IgE
(D) IgG
(E) IgD

375. All of the following are important factors in producing the second higher level of antibody expression **EXCEPT**:

(A) amount of antigen presented
(B) stability of the antigen in the body
(C) surviving sensitized B cells
(D) surviving sensitized T cells
(E) site of the injection

376. Immunoglobulin class switching involves plasma cell clones making

(A) IgG and IgM of different specificities
(B) IgM and IgG of the same specificity
(C) IgM rather than IgG of the same specificity
(D) IgG, IgA or IgE rather that IgM of the same specificity
(E) IgM rather than IgG, IgA or IgE of the same specificity

ANSWERS AND TUTORIAL ON ITEMS 372-376

The answers are: **372-C; 373-B; 374-D; 375-E; 376-D**. The initial humoral response to a specific antigen is usually small since cells of the immune system have no previous exposure to the antigen. IgM is characteristically produced initially followed by an IgG response in the case of serum antibody. A second exposure to the same antigen usually results in a more rapid humoral response with a higher level of IgG versus IgM produced. The IgG level is usually long lived. The difference in humoral responses to the same antigen illustrates immunological memory and the influence of memory cells. Memory cells, either B cells or T cells, are clonally selected cells which have been sensitized by a particular antigen. The residual presence of the antigen in the animal, rather than the location of the exposure is important in maintaining a small population of sensitized cells which, when re-exposed to the same antigen, quickly amplify and produce immunoglobulins. Memory cells are essential in providing natural immunity against reinfections by the same pathogen.

Antigenic switching is the consequence of chromosomal rearrangements in B cells yielding clones which convert from producing specific IgM into producing either IgG, IgA or IgE, with the same specificity. DNA rearrangements and RNA splicing results in the positioning of the variable region of the heavy chain of IgM onto the constant region of the IgG heavy chain. IL-2, IL-4 and IL-10 are important signals in determining which immunoglobulin will be expressed by activated B cells.

Items 377-379

A small population comprising 5-10% of peripheral lymphocytes are called natural killer (NK) cells. They are among the first lymphocytes to attack virus-infected cells and tumor cells.

377. Natural killer cells

(A) are MHC restricted
(B) are thymus-derived
(C) secrete cytotoxins like cytotoxic T cells
(D) require prior sensitization
(E) require antibody for killing

378. Perforins are

(A) proteins
(B) endonucleases
(C) polysaccharides
(D) lipoproteins
(E) ionophores

379. Perforins kill target cells by

(A) making small holes in the cell membrane
(B) degrading chromosomal DNA
(C) degrading ribosomal subunits
(D) disrupting cytoplasmic membranes
(E) degrading messenger RNA

ANSWERS AND TUTORIAL ON ITEMS 377-379

The answers are: **377-C; 378-A; 379-A**. Natural killer (NK) cells comprise a small (5-10%) proportion of peripheral lymphocytes. They respond quickly to either virus-infected or tumor cells. NK cells do not pass through the thymus to mature and do not contain the $\alpha\beta$ T cell receptor. Killing by NK cells is mediated through the contact release of perforins into the membrane of target cells. Perforins are proteins capable of polymerizing into small ionic pores which cause osmotic imbalance and lysis of target cells, but not the producer (NK) cell. The mechanism is similar to that exhibited by cytotoxic T cells. Both cell types produce a protecting protein which prevents perforin self-lysis.

The killing action of NK cells is accentuated by antibody, i.e. antibody-dependent cell cytotoxicity. Cell killing can also occur in the absence of antibody. IL-2 activates NK cells.

Items 380-384

Identical twins were placed under the care of different pediatricians. Each child was vaccinated with a vaccine designed to protect against infections by the polioviruses. One child received multiple doses of the Salk (killed) vaccine while the other received the Sabin (attenuated) vaccine at the times indicated by the arrows. A comparison of different immunoglobulin levels against the vaccine antigens is illustrated in Figure 3.3 below.

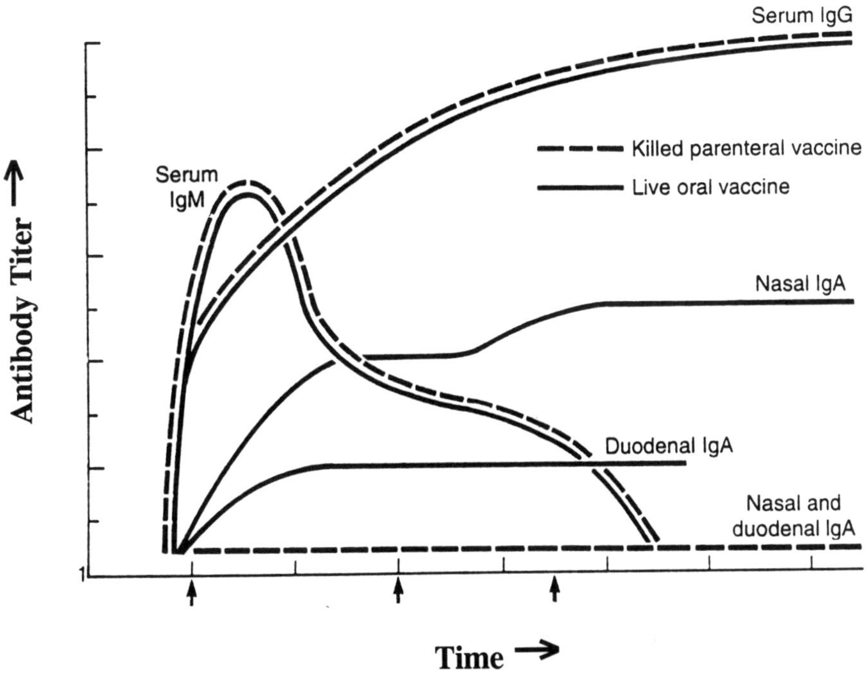

Figure 3.3

380. A primary difference between the two vaccines is the

(A) requirement of booster shots
(B) establishment of mucosal immunity
(C) levels of serum IgG
(D) levels of early IgM response
(E) time course in IgM expression

381. All of the following are correct regarding attenuated vaccines **EXCEPT**:

(A) They confer protective serum IgG levels
(B) They require minimal antigen inoculum
(C) They are indicated for use in immunocompromised individuals
(D) They may revert to virulent virus

382. All of the following are correct regarding killed vaccines **EXCEPT**:

(A) Immunity is shorter-lived than in individuals vaccinated with attenuated vaccines
(B) Excreted viruses immunize people contacted by vaccinated child
(C) Multiple booster injections generally are required
(D) They are safe for immunocompromised individuals
(E) Bacterial contaminants are easily eliminated

383. All of the following are correct regarding advantages of recombinant vaccines **EXCEPT**:

(A) They mimic killed vaccines
(B) They provide improved safety from bacterial contaminations
(C) They eliminate propagation of the pathogen for antigen isolation
(D) They allow smaller and specific B cell or T cell epitope selection
(E) They induce seroconversion in 100% of vaccinated people

384. All of the following are correct regarding attenuated viruses used in vaccines **EXCEPT**:

(A) multiple immunogens may be administered simultaneously
(B) specific vaccines may stimulate mucosal immunity
(C) the same vector may be repeatedly used to deliver different antigens
(D) revertants are possible
(E) vaccines may not be safe for immunocompromised individuals

ANSWERS AND TUTORIAL ON ITEMS 380-384

The answers are: **380-B; 381-C; 382-B; 383-E; 384-C**. Killed vaccines stimulate a humoral response leading to the expression of increased serum IgG levels. The localized presence of the sensitizing antigen triggers the response after antigen is processed. Attenuated vaccines evoke these and additional responses. The initial localized avirulent agent replicates in susceptible host cells and may be transported to a variety of distant organs and tissues. The spread of the organism mimics the natural infection but not the disease. Immunization with attenuated vaccines triggers an IgA response in most cases. Mucosal surfaces become infected before the immune system clears the infection. Killed and attenuated vaccines work best with multiple vaccinations or booster injections. Some individuals require a higher antigenic load for effective vaccination.

Individuals vaccinated with the Sabin vaccine spread avirulent viruses to family members through normal contact. This mode of spread is an advantage of attenuated vaccines, however there are disadvantages. Any revertant organism which regains virulence through a back mutation, recombination with related organisms or other genetic mechanisms, theoretically can infect susceptible individuals and cause disease. This occurs at a very low frequency with the Sabin vaccine but not with the killed Salk vaccine.

The new recombinant vaccines are produced by inserting the genes encoding viral antigens into bacteria or yeast. The host organisms are grown in large batches under conditions favoring viral antigen production. Next, the viral antigens are purified for vaccine use. A new hepatitis B virus vaccine utilizes viral surface antigen produced in yeast.

The latest generation of attenuated vaccines consists of antigen genes inserted into large, attenuated animal viruses such as vaccinia, herpes simplex or RNA tumor viruses. The larger viruses are capable of carrying numerous additional genes and therefore express new antigens which would immunize the recipient. Vaccinia virus can carry as many as 5-10 additional new genes. A limitation of this approach may be the limited usefulness of the virus vector. An individual vaccinated with a vaccinia virus vector would not easily accept a vaccination delivered with the same vector since a high level of antibody against vaccinia, from the first vaccination, would prevent establishment of the second infection and antigen expression. The use of viruses as vaccine vectors is in the developmental stage.

Items 385-389

Choose the best response to each of the items concerning immunoglobulin structure and function.

385. The IgG antibody molecule treated with papain will yield

(A) free amino acids
(B) Fab and Fc fragments
(C) $F(ab')_2$ fragments
(D) heavy and light chains
(E) variable and constant domain fragments

386. Antibody prepared against the γ heavy chain would react with

(A) IgG
(B) IgM
(C) IgA
(D) IgD
(E) IgE

387. Antibody prepared against the ϵ heavy chain will react with

(A) IgG
(B) IgM
(C) IgA
(D) IgD
(E) IgE

388. Antibody prepared against the J chain will react with

(A) IgG
(B) IgM
(C) IgA
(D) IgM and IgA
(E) IgG and IgM

389. Antibody against the following will react against all immunoglobulin classes:

(A) γ heavy chain
(B) κ light chain
(C) λ light chain
(D) α heavy chain
(E) J chain

ANSWERS AND TUTORIAL ON ITEMS 385-389

The answers are: **385-B; 386-A; 387-E; 388-D; 389-C**. The capacity of an individual to produce 1-10 million different antibody specificities resides in the genetic organization and expression of the immunoglobin genes. Once formed, antibodies perform many roles such as lysis of pathogens, activation of B cells, opsonization of pathogens and stimulation of different lymphocyte populations. The structure and overall antigenic composition of the immunoglobulins have both shared and unique properties.

The IgG molecule consists of two heavy chains (MW = 50,000-70,000) and two light chains (MW = 25,000) held together by disulfide bonds. The location of the interchain disulfide bonds and the location of proteolysis sites define the antibody binding portion of the molecule, Fab, and complement binding portion, Fc, fragments. Papain recognizes unique sites on the heavy chain , which once cut, yields two Fab fragments and one Fc fragment. Both types of fragments have polypeptides held together by interchain disulfide bonds. Pepsin cuts the heavy chain at a different locations near the hinge region, yielding connected Fab fragments designated $F(ab')_2$.

A distinguishing feature of the immunoglobulins is their heavy chains. IgG, IgA, IgM, IgD and IgE have unique heavy chains designated γ, α, μ, δ and ϵ, respectively. Therefore anti-β antibody will react with IgG and anti-ϵ antibody will react with IgE. The γ and α heavy chains have subclasses. The J chain peptide is found only in IgA and IgM, therefore antibody against it will react with both. Since nearly all immunoglobulins may contain either λ (lambda) or κ (kappa), light chain antibody to either would react with all immunoglobulin classes. Each individual has genes which encode for either. Put another way, every individual has genes for all isotypes; however, only one is expressed in a given B cell clone.

Items 390-393

A 28 year-old Caucasian woman had been diagnosed with tuberculosis based on a positive tuberculin skin test and the presence of acid-fast bacilli in her sputum. Chest X-rays were negative. She was prescribed long-term antibiotic treatment. During her next annual physical examination, the doctor noticed a faint facial rash. Tests of her CD4+/CD8+ lymphocytes were near normal. Serological test for measles was negative, but positive for certain autoantibodies.

390. The presence of anti-DNA antibody can be used in a preliminary diagnosis of

(A) chronic thyroiditis
(B) allergic encephalitis
(C) rheumatic fever
(D) myasthenia gravis
(E) systemic lupus erythematosus

391. The autoantibodies present in the young woman would most likely react specifically against

(A) thyroid antigens
(B) brain myelin
(C) heart muscle sarcolemma proteins
(D) acetylcholine receptor proteins
(E) H2A-H2B histone proteins

392. Drug-induced autoimmune disease has been observed following long-term treatment with

(A) rifampin
(B) ethambutol
(C) isoniazid
(D) pyrazinamide
(E) streptomycin

393. Management of the disease is done through the use of

(A) gamma interferon (INF-γ)
(B) glucocorticoids
(C) plasmapheresis
(D) irradiation
(E) bone marrow transplants

ANSWERS AND TUTORIAL ON ITEMS 390-393

The answers are: **390-E; 391-E; 392-C; 393-B**. Patients exposed to certain drugs experience a higher probability of developing an autoimmune disease. Patients undergoing treatment for tuberculosis with the drug isoniazid show an increased incidence of systemic lupus erythematosus (SLE). Other drugs such as procainamide, acebutolol and penicillamine have also been associated with a higher frequency of SLE. The target of the autoantibodies is similar even though the drugs differ. The underlying reasons for the effect are not known. Not all drug-induced autoimmune responses develop into SLE.

Patients with SLE may experience a transient facial flush or rash prior to the development of autoantibodies. The antibodies react with the H2A/H2B histone proteins which serve to organize eucaryotic chromosomal nucleic acid into nucleosomes, the fundamental structural subunit of the chromosome. Advanced cases of SLE may result in damage to the kidneys or the central nervous system. There is no cure for SLE, but the disease has been successfully managed with glucocorticoids.

There are many different autoimmune diseases which result in autoantibodies of different specificities. Examples of other common autoimmune disease and the targeted antigens are shown below.

Autoimmune Disease	Target Antigens
chronic thyroiditis	thyroid antigens
allergic encephalitis	brain myelin
rheumatic fever	heart muscle sarcolemma
myasthenia gravis	acetylcholine receptor

Items 394-398

A 45 year-old woman was diagnosed as having a B cell lymphoma of the non-Hodgkin's type. Immunophenotypic studies showed tumor cells bearing both heavy and light chains of the IgM κ class. Attempts at arresting the lymphoma has unsuccessful. The doctors decided to attempt anti-idiotypic antibody therapy to treat her cancer.

394. A feature of this type of B cell lymphoma which made this approach feasible is the

(A) heterogenous population of cancer cells
(B) small number of tumor cells present
(C) impaired immune state of the patient
(D) age of the patient
(E) homogeneous population of cancer cells

395. The cancer cells express the patient's IgM attached to the cell at

(A) the variable region
(B) the hinge region
(C) the Fc region
(D) the constant region of the light chain
(E) many different regions

396. The immunoglobulin idiotype was prepared using a hybridoma system in which the cancer cells were fused to another cell line and clonally selected. The purpose of such an approach would be to yield

(A) a larger number of cancer cells
(B) immunoglobulin of specific idiotype
(C) immunoglobulin of specific anti-idiotype
(D) polyclonal antibodies
(E) cells lacking immunoglobulin markers

397. Hyperimmune anti-idiotypic antibody prepared against the cancer immunoglobulin would attach at

(A) the antigen recognition site
(B) anywhere along the Fab portion of the molecule
(C) anywhere along the Fc portion of the molecule
(D) all surface proteins of the cancer cells
(E) only constant domains of the IgM

398. Preparation of the final "vaccine" incorporated the use of an immunogenic adjuvant to

(A) stabilize the vaccine
(B) make the antigen immunogenic
(C) neutralize the vaccine
(D) inactivate macrophages
(E) prevent an inflammatory response

ANSWERS AND TUTORIAL ON ITEMS 394-398

The answers are: **394-E; 395-C; 396-B; 397-A; 398-B**. A new development used in the treatment of certain types of cancer employs the stimulation of the patient's immune system to produce a stronger immune response against the cancer. The approach varies with different cancers. In the case of B cell lymphomas of the type presented in this case, advantage can be taken of the homogeneous cell type and unique surface marker present of all of the cancer cells. Each cell expressed complete IgM, attached by the Fc region, of the same idiotype.

Fusion of the cancer cells with a hybridoma, of known phenotype, is done to produce large quantities of the same immunoglobulin, IgM, expressed on the cancer cells, but in a soluble form. Clonal selection of the fusion product enables researchers to choose a monoclone expressing the desired complete immunoglobin which is then used as a vaccine antigen. It is made immunogenic by coupling it to a carrier and mixing it with an appropriate adjuvant before injection into the patient. Adjuvants used in humans have included aluminum hydroxide or lipids. The adjuvant stimulates B cells to produce anti-idiotypic antibody in the patient. The desired end result is the production of an immune response, by anti-idiotypic antibodies, against the specific cancer cells which carry the IgM idiotype marker. Presently, this approach in cancer treatment is labor intensive since each vaccine has to be customized for each patient.

Items 399-407

Different tests are used to study the immunological homeostasis and responses to disease. These tests rely on the unique antigenic properties of the organism coupled with a means of visualizing the end results. Match the following tests with the most appropriate basis on which they function in the items below.

(A) Precipitin tests
(B) Agglutination tests
(C) Enzyme-linked immunosorbent assay (ELISA)
(D) Radioimmunoassay
(E) Electron microscopic precipitation
(F) Fluorescent assay
(G) Lymphocyte cytotoxic assay
(H) Complement fixation assay
(I) Western blot
(J) Northern blot
(K) SDS polyacrylamide gel electrophoresis
(L) Hemagglutination assay
(M) Hemagglutination inhibition assay
(N) Polymerase chain reaction (PCR)
(O) Radial immunodiffusion assay
(P) Double immunodiffusion assay
(Q) Dot blot assay
(R) Mixed lymphocyte assay
(S) Skin test

399. Antibody-antigen reaction measured by bound complement

400. MHC relatedness based on DNA synthesis induction

401. Macromolecular separation based on molecular weight

402. Tissue typing based on cytotoxicity to specific immunoglobulins

403. Indirect visual identification or localization of antigen at microscopic level

404. Quantitative and qualitative antigen\antibody determination based on indicator enzyme-substrate reaction

405. Direct antibody-antigen reaction yielding visible particulate product

406. Detection of nucleotide sequence relatedness

407. Delayed type hypersensitivity reaction to an antigen

ANSWERS AND TUTORIAL ON ITEMS 399-407

The answers are: **399-H; 400-R; 401-K; 402-G; 403-F; 404-C; 405-A; 406-N; 407-S**.The measurement of antigen-antibody reactions takes several forms depending on the nature and concentration of the antigen and antibody. Soluble antigens may cross-link with antibody to yield a visible precipitate (A) if the antibody is divalent or monovalent. Particulate antigens may similarly form agglutination (B) reaction products such as in the agglutination of sheep red blood cells or antigen-coated latex spheres by an antibody. An excess of either antigen or antibody decreases the reaction product. A fluorescent assay (F) is useful in the detection of antigen by indirect binding with antibody conjugated with a dye which fluoresces at a particular wavelength. Newer fluorescent systems may incorporate as many a three different conjugated antibodies in the same specimen. Antibody-antigen reactions which fix or bind complement (H) enable precise measurement of reactions which may not readily yield a precipitate. A secondary reaction, such as the complement-dependent lysis of sheep red cells, is used to indicate whether the primary antigen-antibody reaction occurred. The hemagglutination assay (L) measures the presence of hemagglutinin antigen. Many enveloped viruses, such as measles and influenza viruses, display an attachment antigen which also agglutinates sheep red blood cells. Antibody which inhibits the virus-red cell agglutination is the basis of the hemagglutination inhibition assay (M). A certain quantitative dilution, or titer, level of antibody presence is diagnostic. Transmission electron microscopic precipitation (E) of an antigen was first applied to the detection of the hepatitis B virus. The antigen-antibody reaction was not observable at the light or fluorescent microscopic wavelengths.

The double immunodiffusion (P) (Ouchterlony) plate measures antigen-antibody specificity through the formation of precipitin bands where the reactants intersect. Complete, partial or nonhomology are indicated by continuous, spur or crossing precipitin bands, respectively. Radial immunodiffusion plates (O) incorporate a specific antibody, for example human anti-IgA, in the agar, into which serum is added in a well. As the serum diffuses radially IgA, acting now as antigen, is precipitated. The size of precipitin ring is quantitated against reference standards.

Minute concentrations of antibody labeled with a radioisotope, are detectable in a scintillation counting device using radioimmunoassay (D). Antibody coupled with enzymes such as alkaline phosphatase or peroxidase will react with substrates to yield a visible product in enzyme-linked immunosorbent assays (ELISA) (C). The ELISA and radioimmunoassay have both qualitative and quantitative applications.

SDS polyacrylamide gel electrophoresis (K) has been applied to separate proteins by molecular weight. The separated antigen, transferred to nitrocellulose or nylon filters sheets, when treated with antibody, forms the basis of the Western blot (I) used to show the epitope specificity of a particular antibody. The Western blot has been essential in confirmation of AIDS sera shown positive by the ELISA. Northern blots (J) are used to detect separated nucleic acids with tagged, complementary, single-stranded DNA sequences. Southern blots detect RNA in a similar fashion. The nucleic acids for Northern and Southern blots are separated on agarose gels rather than polyacrylamide gels.

The dot blot assay (Q) quantitatively measures RNA sequence relatedness and presence under stringent hybridization conditions. Autoradiographs of exposed X-ray film record the reaction. The polymerase chain reaction (PCR) (N) requires a knowledge of nucleotide sequence information and primer specificity. PCR enzymatically amplifies DNA sequences located between the primer sequences. The final DNA is then isolated by agarose gel electrophoresis and stained with ethidium bromide. DNA sequences of tumor antigens, bacterial, viral and protozoal pathogens have been detected by PCR techniques.

Major histocompatibility complex (MHC) antigens of lymphocytes are detectable by cytotoxic properties of specific antisera (G). Lymphocytes are treated with known antibody and the reaction specificity is measured by killing of test cells indicated by loss of ability to exclude vital dyes. The mixed lymphocyte assay (R) measures MHC relatedness directly by mixing two different lymphocyte cell populations together and monitoring the stimulation of DNA synthesis by radioactive thymidine incorporation, indicating nonrelatedness.

The skin test (S) indicates a delayed type hypersensitivity reaction against an antigen administered intradermally. The size of the resulting induration is measured. Individuals sensitized to a particular antigen will develop a pronounced inflammatory response. The purified protein derivative (PPD) test for Mycobacterium tuberculosis is the most famous and routinely administered skin test.

CHAPTER IV

PARASITOLOGY AND MYCOLOGY

Items 408-412

A 34 year-old female lunched now and then at upscale restaurants in downtown Washington, D.C. At the end of August she experienced severe muscle and joint pain, bilateral periorbital edema, nausea, fever and myalgia. Her differential blood count revealed a marked eosinophilia. A thorough clinical history revealed her frequent consumption of smoked pork. The possibility of a parasitic infection was investigated.

408. The etiologic agent is most probably

(A) Trichinella spiralis
(B) Schistosoma haematobium
(C) Necator americanus
(D) Taenia saginata
(E) Wuchereria bancrofti

409. A confirmatory diagnosis could be established by

(A) examining a fecal specimen for ova
(B) staining a thin blood smear
(C) microscopic examination of skeletal muscle biopsy
(D) performing duodenal aspiration
(E) examining urine sediment for ova

410. A serological test used in the diagnosis of the disease is the

(A) India ink smear
(B) bentonite flocculation test
(C) iodine stain
(D) Giemsa smear
(E) microagglutination

411. Spread of the causative pathogen within the patient occurs through

(A) association with red blood cells
(B) association with macrophages
(C) the lymphatics
(D) the bloodstream
(E) the respiratory tract

412. Prevention of the disease is best accomplished through

(A) vaccination
(B) passive immunization
(C) proper hygiene practices
(D) vector control measures
(E) proper cooking of meats

ANSWERS AND TUTORIAL ON ITEMS 408-412

The answers are: **408-A; 409-C; 410-B; 411-D; 412-E**. This is classical case of trichinosis caused by the nematode Trichinella spiralis. The infecting larvae are acquired by humans from the consumption of inadequately cooked pork or wild carnivorous animals. The larvae penetrate the mucosa of the small intestine and then become adults there. Each fertilized female releases hundreds of larvae into the mucosa. These then enter the bloodstream from which they penetrate and lodge in many tissues. Most larvae eventually die; however, those reaching skeletal muscle survive for 5-10 years. Within skeletal muscle, the larvae encyst and calcify, causing the characteristic muscle pain and fever. The oculomotor, intrinsic lingual, deltoid, pectoral and diaphragmatic muscles and others are most often infected. Transmission of the larvae occurs when infected muscles and cysts are eaten. The digestive enzymes of the consuming animal then free the larvae which mate and initiate new cycles of infection. Most infections are asymptomatic. The proper cooking of meats, especially pork, will prevent the spread of the parasite. The basic life cycle of Trichinella spiralis is shown in Figure 4.1.

Diagnosis is confirmed through the observation of the small (1.5 mm) larvae within muscle biopsy. A serological test, the bentonite flocculation test, detects specific antibody in a slide agglutination test consisting of antigen-coated bentonite particles. Enzyme-linked immunosorbent assays have also been developed.

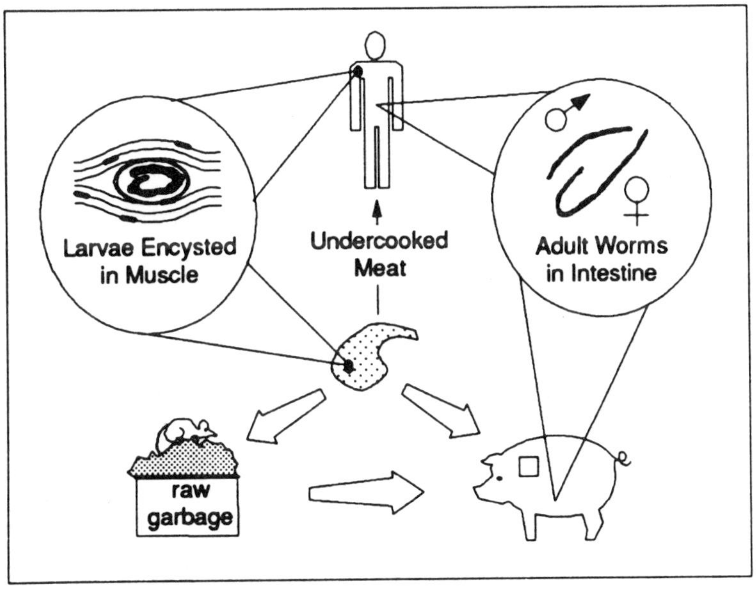

Figure 4.1

Items 413-417

Eight months after returning from a month-long vacation to rural Mexico, a 25 year-old man walked into his physician's office and handed him a jar containing a 1x2 inch long flat object which he had retrieved from the surface of his stool. The physician recognized the object as a proglottid. On microscopic examination, the proglottid had 8 uterine branches. The astute physician correlated this patient's episodes of severe headaches and disequilibrium to the larval stage of this parasite.

413. The physician would correctly diagnose an infection caused by

(A) Taenia solium
(B) Echinococcus granulosus
(C) Enterobius vermicularis
(D) Taenia saginata
(E) Diphyllobothrium latum

414. Examination of the rounded head of the adult parasite would reveal

(A) 4 suckers
(B) 4 suckers and a circle of hooks
(C) 6 suckers and a circle of hooks
(D) 6 suckers
(E) 5 suckers and a circle of hooks

415. The first drug of choice prescribed to treat the adult parasite would be

(A) pyrantel pamoate
(B) chloroquine
(C) diiodohydroxyquin
(D) niclosamide
(E) praziquantel

416. The infection is acquired by

(A) drinking water containing rhabditiform larvae
(B) ingesting contaminated raw or rare pork
(C) penetration of the skin by filariform larvae
(D) ingesting contaminated rare or raw beef
(E) inhalation of dried bird feces

417. The form of the parasite first encountered most frequently by humans is the

- (A) proglottid
- (B) egg
- (C) larva
- (D) amoeba
- (E) scolex

ANSWERS AND TUTORIAL ON ITEMS 413-417

The answers are: **413-A; 414-B; 415-D; 416-B; 417-C**. There are two classes of flat worms of significant human disease importance; Cestoda (tapeworms) and Trematoda (flukes). The patient in this instance has a case of tapeworms caused by Taenia solium. Its natural host are swine. Taenia solium is distinguishable from its bovine counterpart, Taenia saginata, by the morphology of the proglottid. The former has 5-10 uterine branches in the end proglottid versus 15-20 found in the latter. The head, or scolex of Taenia solium, found attached to the intestines has 4 suckers and a circle of hooks which hold the parasite in place. The scolex of T. saginata also has 4 suckers, but lacks the hooks (Figure 4.2 on next page).

Diphyllobothrium latum (fish tapeworm), has 2 elongated suckers on the scolex, while Echinococcus granulosus (dog tapeworm) is relatively short with only 3 proglottids. Enterobius vermicularis causes pinworm infections in humans.

Infection usually occurs through the ingestion of larvae, also called cysticerci, which attach to the small intestine, and develop into adult forms which may exceed 4-5 meters in length. The plate-like proglottids produce eggs which are passed in the feces along with end proglottids. The eggs normally reinfect pigs and develop into larvae which localize in the skeletal muscles of the animal. The larvae are consumed in raw or rare pork, thus initiating a cycle of infection. The human consumption of eggs in food or water contaminated by feces causes a more profound disease in humans called cysticercosis. In this infection, the larvae hatch and gain access to many organs including the eyes or brain.

Niclosamide is the first drug of choice. Praziquantel is sometimes used as a last resort. The parasite can also be surgically removed if drugs are ineffective.

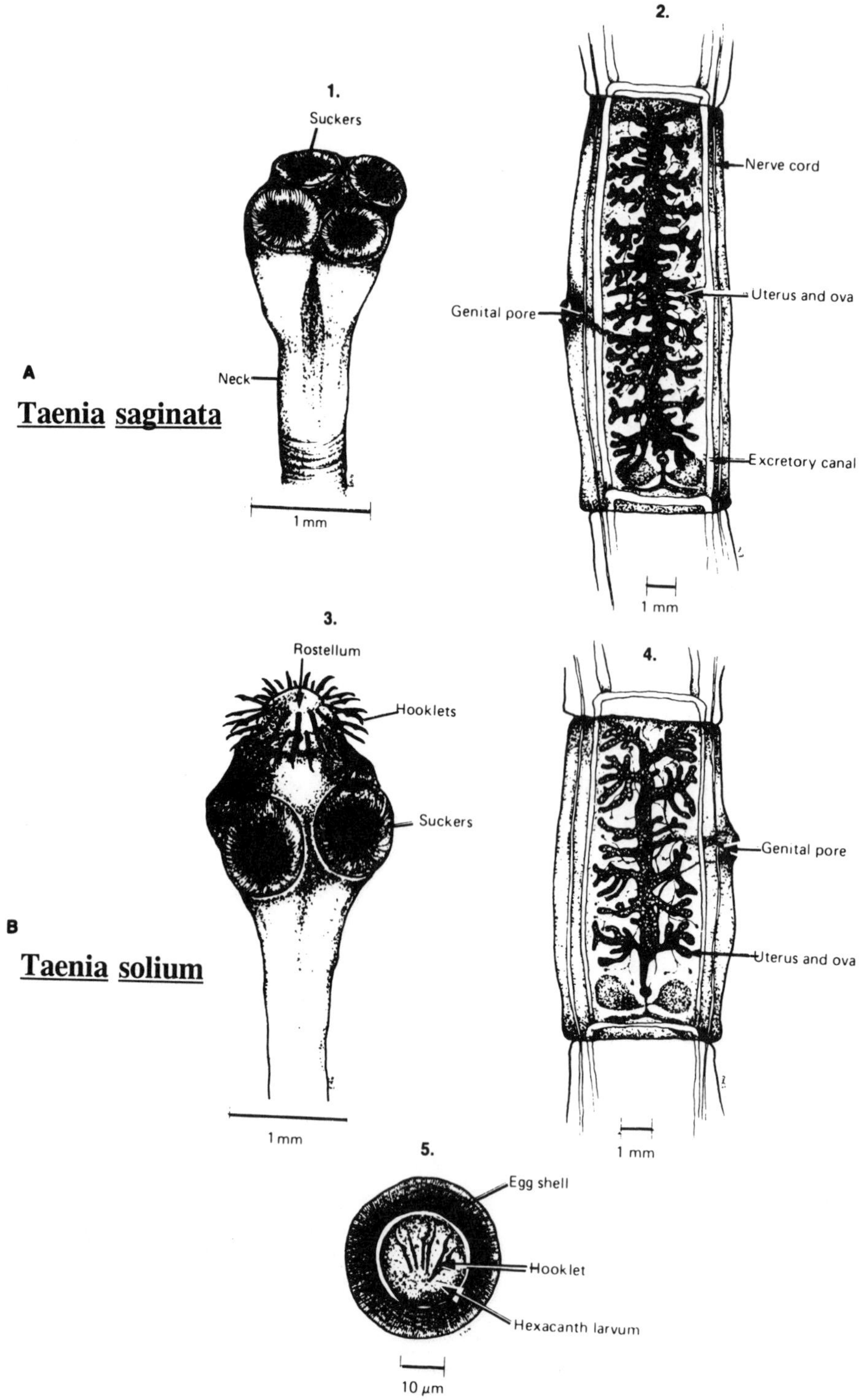
A
Taenia saginata
1.
Suckers
Neck
1 mm
2.
Nerve cord
Uterus and ova
Genital pore
Excretory canal
1 mm
B
Taenia solium
3.
Rostellum
Hooklets
Suckers
1 mm
4.
Genital pore
Uterus and ova
1 mm
5.
Egg shell
Hooklet
Hexacanth larvum
10 μm

Figure 4.2

Items 418-424

A 25 year-old Cambodian refugee in her first trimester of pregnancy developed abnormal symptoms consisting of frequent high fever (104°F) accompanied by malaise, headache, confusion and nausea in three-day cycles. Initially, she attributed the symptoms to her pregnancy but later reported them to her physician. Clinical findings included thrombocytopenia, tachycardia, splenomegaly, hemoglobinuria and anemia. She was admitted to the hospital and shortly went into a coma.

418. The correct presumptive diagnosis would be an infection caused by

(A) Plasmodium vivax
(B) Plasmodium falciparum
(C) Toxoplasma gondii
(D) Pneumocystis carinii
(E) Plasmodium ovale

419. A confirmation of the diagnosis could be made by

(A) a Giemsa-stained thin blood smear
(B) an indirect fluorescent antibody test (IFA)
(C) culturing a blood sample in an enriched medium
(D) direct microscopic examination of a fresh blood specimen
(E) patient history

420. Administration of the following drug(s) of choice would normally result in a dramatic cure.

(A) quinine sulfate, pyrimethamine and sulfadiazine
(B) chloroquine phosphate
(C) pyrimethamine, sulfadoxine and chloroquine
(D) quinine dihydrochloride, i.v.
(E) mefloquine

421. Organisms resistant to the first-choice treatment are given

(A) quinine sulfate, pyrimethamine and sulfadiazine
(B) chloroquine phosphate
(C) pyrimethamine, sulfadoxine and chloroquine
(D) quinine dihydrochloride, i.v.
(E) mefloquine

422. The vector for transmission of this disease is

(A) mosquitos
(B) fleas
(C) gnats
(D) raccoons
(E) parrots

423. The developmental form of the parasite transmitted during the biting process is the

(A) sporozoite
(B) merozoite
(C) hypnozoite
(D) macrogametes
(E) microgametes

424. Microscopic examination of stained, infected red cells would reveal

(A) ring-shaped tropozoites
(B) spherical gametocytes
(C) crescent-shaped gametocytes
(D) rod-shaped gametocytes
(E) beaded gametocytes

ANSWERS AND TUTORIAL ON ITEMS 418-424

The answers are: **418-B; 419-A; 420-B; 421-A; 422-A; 423-A; 424-C**. This woman is suffering from malaria caused by an infection by Plasmodium falciparum. Malaria affects over 200 million people world wide. Endemic regions include tropical and subtropical regions of Asia, Africa, South and Central America. Plasmodium falciparum is the most virulent of the four species which cause the disease. P. vivax, P. ovale, and P. malariae are the other members of the group. Fatalities are less frequent in cases involving P. vivax and P. ovale. Both produce recurrent episodes of latent infections.

Initial transmission of the disease begins with the injection of sporozoites from the mosquito vector. Anopheles, Aedes, and Culex mosquitos may transmit the disease. Individuals infected with P. falciparum have pronounced parasitemia in all stages of red cell maturity and a higher level of parasitemia than other malaria organisms. Identification of P. falciparum is obtained from the examination of Giemsa-stained blood smears. Crescent-shaped gametocytes are present in many red blood cells.

The drug of choice in treating malaria is chloroquine. Resistant organisms are treated with a combination of drugs consisting of quinine sulfate, pyrimethamine and sulfadiazine. New malaria vaccines which incorporate antigen incorporated into liposomes have been used in limited human trials.

Items 425-429

A 32 year-old office worker at the Venezuelan embassy was admitted to The George Washington University Hospital following a grand mal seizure. During his youth he had experienced seizures which included numbness of the upper left extremity, unconsciousness and residual left-side headaches. He had not experienced such symptoms for many years until they resurfaced several months ago when he recalled similar progressive weakness, sharp pain in his upper left arm, blurred vision and motor disturbances. An encephalogram revealed calcification in the 3rd ventricle.

425. The history and symptoms are consistent with a preliminary diagnosis of

(A) amebiasis
(B) cysticercosis
(C) malaria
(D) schistosomiasis
(E) pin worms

426. This disease is caused by which of the following parasites

(A) Taenia solium
(B) Echinococcus granulosus
(C) Enterobius vermicularis
(D) Taenia saginata
(E) Diphyllobothrium latum

427. He probably acquired the infection by ingesting

(A) zoitocysts in undercooked beef
(B) cysts in drinking water
(C) cercariae from stream water
(D) parasite eggs from contaminated food or water
(E) adult parasites

428. An effective drug useful in the treatment of this disease is

(A) praziquantel
(B) metronidazole
(C) primaquine
(D) pyrimethamine
(E) amphotericin B

429. A closely related organism but which causes neither cysticercosis nor pronounced human infections is

(A) Taenia solium
(B) Echinococcus granulosus
(C) Enterobius vermicularis
(D) Taenia saginata
(E) Diphyllobothrium latum

Items 430-433

A young woman in southern Florida desired a fashionable suntan. The results were not what she had anticipated. Her tan appeared uneven with well outlined, hypopigmented lesions on her upper torso. Her dermatologist performed a microscopic examination of skin scales using potassium hydroxide solution and observed yeast cells and hyphae resembling "spaghetti and meat balls".

430. The history and examination are consistent with a preliminary diagnosis of

(A) ringworm
(B) tinea versicolor
(C) tinea unguium
(D) sporotrichosis
(E) candidiasis

431. Which is the most appropriate nutrient media for culturing this organism?

(A) Lowenstein-Jensen medium
(B) Sabouraud's agar
(C) nutrient blood agar
(D) Mueller Hinton medium
(E) Conradi Drigalski agar

432. An antifungal compound of choice for treating the infection is

(A) tolnaftate
(B) undecylenic acid
(C) zinc oxide
(D) miconazole
(E) salicylic acid

433. The organism responsible for the infection is

(A) <u>Cladosporium</u> <u>werneckii</u>
(B) <u>Piedraia</u> <u>hortae</u>
(C) <u>Sporothrix</u> <u>schenckii</u>
(D) <u>Pityrosporum</u> <u>orbiculare</u>
(E) <u>Trichosporon</u> <u>cutaneum</u>

ANSWERS AND TUTORIAL ON ITEMS 425-429

The answers are: **425-B; 426-A; 427-D; 428-A; 429-D**. The patient is suffering from cysticercosis in the brain caused by the parasitic tapeworm Taenia solium. Infection occurs through the ingestion of parasite eggs in foods or drinking water contaminated with human feces. The ingested eggs hatch in the small intestines where the embryos, called oncospheres, enter into the blood stream and spread to all parts of the body including the brain. In the brain, the parasites encyst, form cysticerci and calcify. The initial infection does not evoke an inflammatory response. As the parasites die, an inflammatory response is induced.

Massive doses of praziquantel are effective in treating cysticercosis in the brain. Surgical removal of the cysticercus may be necessary to eradicate the organism.

Individuals infected with the closely related tapeworm Taenia saginata can not get cysticercosis. This parasite is acquired from the ingestion of contaminated beef. Most infections are asymptomatic and localized in the small intestines.

ANSWERS AND TUTORIAL ON ITEMS 430-433

The answers are: **430-B; 431-B; 432-D; 433-D**. Superficial fungal infections may develop on many different parts of the body. They are not restricted to moist areas. This is most evident in tropical regions of the world. The discrete areas of hypopigmentation in this patient are due to an infection called pityriasis versicolor (tinea versicolor) named for the variations in the pigmentation produced by the causative organism. The disease is caused by Malassezia furfur (Pityrosporum orbiculare). Infections are most common in the tropics. Exposure to sunlight favors growth of the organism resulting in the appearance of a slightly scaly, blotchy area.

Cultivation of the organism on synthetic media is not usually performed for diagnostic purposes. This organism can be cultivated on Sabouraud's agar like many other fungi. Some fungal infections produce pigmentations characteristic of the fungal hyphae. The pigmentation produced by this organism ranges from pink to yellow-brown. A cell scraping from the infected area onto a slide containing a 10% potassium hydroxide solution, which dissolves most exfoliated cells but not the rigid fungal cell walls, aids in the identification of this organism.

Most cutaneous fungal infections are treated topically with creams or ointments. Miconazole is an effective topical antifungal compound used in the treatment of pityriasis versicolor.

Items 434-437

A 68 year-old migrant worker complained to coworkers of annoying foot pain and infection. He finally sought medical help at a local clinic. Examination of the foot showed a large, scaly, crusted, cauliflower-like lesion on his ankle. This lesion resembled a deep, tumorous infection which ruptured and drained to the surface.

434. Infection of this type are characteristic of those caused by

(A) histoplasmosis
(B) blastomycosis
(C) sporotrichosis
(D) chromomycosis due to dematiaceous fungi
(E) eumycotic (fungal) mycetoma

435. The source of the infecting organism was most likely

(A) birds
(B) livestock
(C) snails
(D) soil
(E) airborne

436. Treatment of the primary infection may require

(A) surgical excision
(B) irradiation
(C) antifungal compounds
(D) antibiotics
(E) immune modulators

437. Deep lesions of this type are caused by which of the following genera?

(A) <u>Petriellidium</u>
(B) <u>Malassezia</u>
(C) <u>Cladosporium</u>
(D) <u>Trichophyton</u>
(E) <u>Microsporum</u>

Items 438-441

A 33 year-old woman was diagnosed with an HIV infection two years ago and had undergone aggressive chemotherapy. Her CD4+ T cell count had dropped below 200 cells/mm^3. She developed a pneumonia infection that resulted in the sudden onset of fever, dyspnea, coughs and tachypnea. X-ray images revealed diffuse interstitial pneumonia. Bronchial lavage specimens did not grow bacteria. Other tests ruled out viral pneumonia.

438. The immunocompromised state of the patient makes her a prime candidate for which of the following respiratory pathogens?

(A) Candida albicans
(B) Aspergillus fumigatus
(C) Pneumocystis carinii
(D) Histoplasma capsulatum
(E) Blastomyces dermatitidis

439. Inhalation of the organism into the lungs initially causes

(A) cell necrosis
(B) massive exfoliation of epithelial cells
(C) inflammation and a frothy exudate that blocks respiration
(D) secondary bacterial infections
(E) granuloma formation

440. Laboratory diagnosis is also based on observation of which of the following in silver-stained bronchial lavage specimens?

(A) mycelia
(B) hyphae
(C) cysts
(D) spores
(E) pigmented microcolonies

441. Drugs of choice used in treating the infection include

(A) trimethoprim-sulfamethoxazole
(B) penicillin
(C) amphotericin B
(D) sodium stibogluconate
(E) chloroquine

ANSWERS AND TUTORIAL ON ITEMS 434-437

The answers are: **434-E; 435-D; 436-A; 437-A**. A variety of soil-borne fungi such as Petriellidium, Pseudallescheria and Madurella sp. can cause subcutaneous infections. Fungal eumycotic mycetoma, also called maduromycosis, is initiated by a traumatic wound followed by soil contamination. The organisms produce local swelling and a painless tumorous growth which eventually drains to the surface. The fungi spread through the circulatory system, muscle and bone. Nerve damage is usually minimal. Secondary bacterial contaminations may follow. Infections are most prevalent in tropical and subtropical areas.

Advanced cases of infection usually require surgical excision of the affected area. Mycetomas of bacterial origin, caused by Nocardia, Actinomyces or Streptomyces usually produce more copious amounts of draining discharge along with significant tissue and bone damage.

ANSWERS AND TUTORIAL ON ITEMS 438-441

The answers are: **438-C; 439-C; 440-C; 441-A**. Immunocompromised patients are prime candidates for infection by the fungus Pneumocystis carinii. Cysts are inhaled in to the lungs where the germinate and release 2-8 trophozoites. Each trophozoite eventually develops into a cyst. Antigens produced by the organism initiate an inflammatory response resulting in the production of frothy mucous material which interferes with normal oxygen transport, thus the observed difficulty in breathing.

A bronchial lavage or sputum specimen from a patient usually contains numerous cysts which are easily recognized in silver-stained preparations or Giemsa-stained smears. The cysts are nearly spherical but have flattened sides and usually appear in small clumps.

The fatality rate in untreated Pneumocystis pneumonia approaches 100%. Vigorous therapy with trimethoprim-sulfamethoxazole reduces the rate to 50%. The drugs inhibit parasite dihydrofolate reductase and dihydropteroate synthase activities. Most individuals have been exposed to the fungus but experience only asymptomatic infections. This fungus is profoundly opportunistic in immunocompromised patients.

Items 442-446

A 28 year-old woman with an extensive prior history of IV heroin abuse, experienced severe headaches, disorientation, high fevers, lethargy and mild seizures lasting two weeks. She did not travel outside her home state of California, lived alone but had three cats and a small dog as house pets. Laboratory studies showed the presence of serum antibody against HIV and a low CD4+ T cell count of 120 cells/mm^3. Chest X-rays appeared normal. Cerebrospinal fluid was negative for bacterial infection, negative for antibody against the picornaviruses and normal for protein. A heterophile test was negative. CT scans showed several small lesions in her frontal lobe.

442. The history and immune state of the individual are consistent to support further serological tests for

(A) malaria
(B) psittacosis
(C) Pneumocystis carinii pneumonia
(D) toxoplasmosis
(E) tuberculosis

443. A Giemsa-stained thin blood smear would most likely show

(A) large atypical monocytes
(B) clumped red blood cells
(C) crescent-shaped trophozoites
(D) circular cysts with flattened sides
(E) rod-shaped bacteria

444. Infection was most likely acquired through

(A) sexual contact
(B) IV drug abuse
(C) ingestion of contaminated foods
(D) inhalation of animal feces
(E) an animal bite

445. Drugs of choice used in treating the infection include

(A) trimethoprim-sulfamethoxazole
(B) sulfonamide-pyrimethamine
(C) amphotericin B
(D) sodium stibogluconate
(E) chloroquine

446. Pregnant women should not be exposed to the pathogen because

(A) there is a higher virulence in that population
(B) they are partially immunosuppressed
(C) there is risk of transplacental infection of fetus
(D) there is a greater possibility of CNS involvement
(E) there is a higher frequency of chronic infections

ANSWERS AND TUTORIAL ON ITEMS 442-446

The answers are: **442-D; 443-C; 444-C; 445-B; 446-C**. The patient suffers from toxoplasmosis caused by Toxoplasma gondii. Cats serve as a primary reservoir for these parasites. Infections in humans are incidental, but more pronounced in immunocompromised individuals. In the latter population, especially in HIV positive patients, infections of the CNS are common .

Human infection occurs by the ingestion of foods contaminated by cat feces. Cysts mature and release sporozoites which may enter the blood stream and infect red blood cells. Eventually, the organisms reach the brain and form cysts which induce inflammatory responses and tissue damage. Infections of the eye, such as chorioretinitis, resulting in severe visual impairment, may also result.

Pregnant women infected with the parasite during the third trimester show a high (>60%) probability of passing the parasite to the fetus. Stillbirth or mental retardation may result. If a pregnant woman washes her hands frequently and avoids cat feces, she can greatly reduce the incidence of infections and disease. Treatment for toxoplasmosis includes the use of sulfonamides and pyrimethamine in adults.

Items 447-453

Match the parasitic organism in the answers with the most appropriate clinical scenario in the items below.

(A) Giardia lamblia
(B) Taenia ~~saginata~~ solium <10
(C) Toxoplasma gondii
(D) Taenia ~~solium~~ saginata >10
(E) Entamoeba histolytica
(F) Trichomonas vaginalis
(G) Trypanosoma cruzi
(H) Leishmania mexicana
(I) Pneumocystis carinii
(J) Ascaris lumbricoides
(K) Enterobius vermicularis
(L) Necator americanus
(M) Schistosoma mansoni

447. A 19 year-old homosexual male developed a progressive pneumonia which did not respond to antibiotics. A silver stain of sputum and bronchial lavage specimens showed small circular cysts containing sporozoites.

I

448. An underweight 17 year-old girl from Panama was visiting her cousin in Philadelphia. She experienced fever, diarrhea and severe right upper quadrant pain. A fresh stool specimen revealed mobile trophozoites and cysts with 4 nuclei.

E

449. A 53 year-old camper developed a case of diarrhea and stomach cramps during the second week of a fishing trip in the Colorado mountains. Trophozoites, with owl-like features, and cysts were observed in his stool specimen. A duodenum string test was also positive for trophozoites. A

450. An emaciated 18 year-old female complained of constant headaches and diarrhea. A stool specimen revealed flat proglottids. Microscopic examination showed less than 10 uterine branches per proglottid.

B

451. A 24 year-old female medical student in Washington, D.C. developed an irritating vaginal infection which produced an abundance of watery, whitish discharge. Microscopic examination of a wet mount showed pear-shaped trophozoites.

F

452. A mother noticed that her 9 year-old son was frequently scratching at his anus. The scratching resulted in a mild bacterial infection. Examination of his perianal region using the transparent tape method revealed tiny worms and small eggs.

K

453. A 28 year-old female CIA agent gave birth to an infant with splenomegaly and jaundice. Serological tests for the hepatitis viruses were negative. The mother had not traveled during the previous year. She lived with her boyfriend and three cats. Blood smears of the infant showed crescent-shaped trophozoites.

C

ANSWERS AND TUTORIAL ON ITEMS 447-453

The answers are: **447-I; 448-E; 449-A; 450-B; 451-F; 452-K; 453-C**. Giardia lamblia (A), giardia. It has both trophozoite and cyst forms. The binucleate trophozoite has two basal bodies, is shaped like a sting ray with prominent "owl-like eyes". The organisms are 5-15 μm in wide and 9-21 μm long. Four pairs of long flagella border the body. Each cyst contains two trophozoites.

Taenia solium (B), pork tapeworm. The head, or scolex, has 4 suckers and a circle of hooks which anchor it. The long series of proglottids have 8-12 lateral branches in the uterus. A mature organism may grow as long as 5 meters with hundreds of aligned proglottids. Infected individuals may appear anorexic.

Toxoplasma gondii (C). The elliptical 10 μm oocyst may have 2 sporocysts, each with 4 sporozoites. The crescent-shaped tropozoites, 3x7 μm, have a large nucleus and asymmetric organelle distribution. Cats carry the organism and transmit it to humans. Infected mothers can infect their unborn fetuses, resulting in congenital damage to the brain, liver and/or spleen.

Taenia saginata (D), beef tapeworm. The head has four suckers all on the same plane. There are no circular hooks. The proglottids have 15-25 lateral branches in the uterus. May reach 10 meters in length with several hundred proglottids.

Entamoeba histolytica (E). The circular cyst, 15 μm, have 4 nuclei and 2 cigar-shaped chromatoid bodies. The Trophozoite, 12-20 μm, is elliptical with pseudopod-like projections, and vacuolated. The cysts of Entamoeba coli have 8 nuclei. Transmission occurs through contaminated drinking water.

Trichomonas vaginalis (F). The trophozoite, 7x15 μm, is pear-shaped with a single nucleus, 5 flagella, with one appearing attached to the body. A long rod, called the axostyle, traverses and extends past the body longitudinally. The organism is transmitted through sexual contact.

Trypanosoma sp. (G). These long ,15-30 μm, hemoflagellates differ in shape but most have a single flagellum running longitudinally, a central nucleus and kinetoplast.

Leishmania sp. (H). Amastigote forms are small, 1-2 μm, with a single nucleus and kinetoplast. Many found in cytoplasm of infected white cells. The promastigotes are flagellated.

Pneumocystis carinii (I). The cysts, 5-8 μm, are nearly circular but often have flattened sides, contain 2-8 small sporozoites and may appear with a honeycomb-like surface. The mature trophozoites are pleomorphic with a single eccentric nucleus. This is an opportunistic organism commonly found infecting immunocompromised patients.

Ascaris lumbricoides (J). Largest of the intestinal helminths, 150-350 mm, with a creamy cuticle and blunted anterior end.

Enterobius vermicularis (K), pinworm. Females, 1x10 mm, creamy colored with blunt anterior end. The oval egg is flattened on one side. These organisms are difficult to detect in stool specimens. The transparent tape test captures worms and eggs from the skin of the perianal areas for easy microscopic examination.

Necator americanus (L), hookworm. Adults measure 1x10 mm with blunt anterior head and 'hook' shaped body. Eggs measure 20x60 μm with 4-8 cells visible.

Schistosoma mansoni (M). Eggs appear reddish-brown, 50x90 μm, with a characteristic large lateral spine near the posterior end. A rigid capsule surrounds the egg.

Items 454-457

A family of three on a camping trip went swimming in a fresh water pond while traveling in Maine. Eight days later, the normally healthy 3 year-old daughter developed continuous headaches, fever and increasing lethargy. The physician noted she was unresponsive to painful stimuli. A CSF specimen was negative for bacteria but high in protein level. A CT scan revealed massive encephalitis. Four days after admission to the hospital she died.

454. The symptoms and history of the child's illness are consistent with a preliminary diagnosis of meningoencephalitis caused by

(A) Babesia microti
(B) Entamoeba histolytica
(C) Naegleria fowleri
(D) Acanthamoeba
(E) Balantidium coli

455. Exposure to the causative agent occurred most likely through

(A) contact with domestic pets
(B) inhalation of bird feces
(C) swallowing and inhaling of contaminated fresh water
(D) ingestion of contaminated foods
(E) soil contaminated cuts or bruises

456. An effective drug for this disease, if administered early and aggressively, is

(A) praziquantel
(B) niclosamide
(C) amphotericin B
(D) quinine
(E) metronidazole

457. The causative agent is known to infect and cause disease in

(A) primarily young individuals
(B) primarily old individuals
(C) any age group
(D) more males than females
(E) more females than males

ANSWERS AND TUTORIAL ON ITEMS 454-457

The answers are: **454-C; 455-C; 456-C; 457-A**. This is a representative case of primary amebic meningoencephalitis (PAM). The two organisms most frequently associated with this type of illness are Naegleria fowleri and Acanthamoeba. PAM infections caused by Naegleria primarily occur in normally healthy, immunocompetent, young individuals; whereas, infections caused by Acanthamoeba occur in older, immunocompromised individuals.

Contact with Naegleria occurs during exposure to contaminated fresh water. Exposure does not ensure infection and disease. Symptoms usually begin within 1-2 weeks of exposure. Nearly all cases are fatal unless an early diagnosis is made. The patient history is essential for an accurate diagnosis. Unfortunately, PAM symptoms resemble bacterial meningoencephalitis symptoms, e.g. high protein levels and white cell infiltration both in CSF. The organisms are not readily observed in Gram-stained CSF. A high speed centrifugal pellet of CSF may contain the amoeba which can then be observed in wet mount preparations or permanent slides. Aggressive treatment with amphotericin B and intensive supportive care has been effective in treating this disease.

Items 458-462

A new bride became concerned when she experienced a vaginal infection which produced an unusual thick, cream-colored, mucous vaginal discharge and vulvar itching. Sexual intercourse with her husband resulted in added discomfort.

458. The organism responsible for the infection is a

(A) bacterium
(B) fungus
(C) virus
(D) protozoan
(E) nematode

459. A Gram stain of the mucous discharge would show

(A) gram-positive cocci
(B) gram-negative cocci
(C) gram-negative rods
(D) gram-positive rods
(E) gram-positive budding forms

460. The organism responsible for the infection is

(A) Entamoeba histolytica
(B) Entamoeba hartmanni
(C) Epidermophyton floccosum
(D) Microsporum audouinii
(E) Candida albicans

461. The infection most likely derived from

(A) sexual transmission from husband
(B) opportunistic infection from normal flora
(C) indirect contact from clothing
(D) non-sexual contact from other individuals
(E) airborne spores

462. All of the following drugs have been shown effective against the causative agent **EXCEPT**:

(A) amphotericin B
(B) nystatin
(C) flucytosine
(D) imidazoles
(E) doxycycline

ANSWERS AND TUTORIAL ON ITEMS 458-462

The answers are: **458-B; 459-E; 460-E; 461-B; 462-E**. The symptoms encountered by the young woman are characteristic of a yeast infection caused by Candida albicans. Many individuals harbor Candida as part of their normal flora. They can be isolated from the oral cavity, lower respiratory tract and female genital tract of healthy asymptomatic individuals. The organisms are kept are bay by the host's immune system. Candida albicans is an opportunistic organism which can cause infections if given access through the body's first line of defense, such as minute abrasions of the vaginal mucosa occurring during normal sexual activity. Changes in vaginal pH can prove favorable for prolific growth of yeast resulting in pronounced infection and tissue damage.

Candida growth is marked by the presence of large, 4-6 μm, oval, gram-positive budding forms. Pseudohyphae are occasionally detected. The copious secretion of thick, cream-colored discharge, itching sensation and abundance of easily identified structures makes diagnosis of a yeast infection relatively easy.

The organisms respond to a variety of antifungal drugs including amphotericin B, nystatin, flucytosine and the imidazoles. Doxycycline, an antibacterial drug, has no effect on Candida albicans.

Items 463-468

A Cub Scout troop went camping in rural Colorado. After a long day of hiking, the boys went swimming in a pool frequently used by several different troops in the area. The next week, four of the boys developed diarrhea. Their stools had an unusual foul smell, a greasy appearance and light density. Their stools appeared fatty but were neither bloody nor mucous. The ill boys also experienced nausea, vomiting and fever.

463. The organism most likely responsible for the illnesses is

(A) Entamoeba histolytica
(B) Giardia lamblia
(C) Trichomonas vaginalis
(D) Shigella dysenteriae
(E) Ascaris lumbricoides

464. The most probable source of infection was exposure to

(A) untreated contaminated water
(B) soil-borne pathogens
(C) aerosolized organisms
(D) insect-borne pathogens
(E) a zoonotic illness

465. The specimen most useful for laboratory diagnostic purposes is

(A) blood
(B) sputum
(C) cerebrospinal fluid
(D) urine
(E) feces

466. The microscopic examination of specimen would reveal

(A) gram-positive budding forms
(B) long 1x20 μm worms
(C) proglottids with 15-20 uterine branches
(D) trophozoites with eye-like structures
(E) gram-negative rod-shaped bacteria

467. The organism is found in which location in infected humans?

- (A) duodenum, jejunum and upper ileum
- (B) only lower ileum
- (C) stomach
- (D) esophagus
- (E) equally dispersed throughout digestive tract

468. An effective drug used in the treatment of the disease is

- (A) amphotericin B
- (B) nystatin
- (C) flucytosine
- (D) imidazole
- (E) furazolidone

ANSWERS AND TUTORIAL ON ITEMS 463-468

The answers are: **463-B; 464-A; 465-E; 466-D; 467-A; 468-E**. Giardia lamblia are found in untreated bodies of water exposed to human or animal feces. Ingestion of the water may lead to infection. The organisms may be transmitted directly via the fecal-oral route or through anal sex. Giardia are among the most frequent causes of diarrheal infections caused by intestinal parasites. The organisms will infect all ages; however, the poor hygiene practices of children render them more prone to Giardia infections. The ingested organisms colonize the duodenum, jejunum and upper ileum.

An examination of the feces normally reveals the presence of cysts and trophozoites. The latter flagellates are tear-drop shaped and exhibit a tumbling motion is wet mount preparations. The two nuclei of the trophozoites give them an owl-like appearance. Two parabasal bodies are located adjacent to the nuclei. Cysts contain two binucleate trophozoites.

The presence of Giardia in the small intestines results in the malabsorption of fats and carbohydrates. This produces fat-laden feces which, when acted upon by intestinal bacteria, produce an unusual foul odor, diarrhea and a cramping sensation. Giardiasis in adults is treated with either quinacrine or metronidazole. Children are treated with furazolidone.

Items 469-473

Three U.S. oil prospectors assisted in the surveying a potentially productive oil field in the jungles of Brazil. One month after returning to the U.S., two of the men developed lesions on their faces, arms and necks which progressively increased in size and began to ulcerate. The men also experienced mild fever, weight loss and reduced appetites. A Giemsa stain of pus taken from the lesions showed white cells and macrophages. Some cells were degenerate with vacuolated cytoplasm containing numerous, 1-2 μm structures with darkly stained nucleoids.

469. The history of the individuals and description of the intracytoplasmic structures observed in infected white cells are consistent with a preliminary diagnosis of

(A) malaria
(B) schistosomiasis
(C) leishmaniasis
(D) giardiasis
(E) sleeping sickness

470. The objects observed in infected white cells are most likely

(A) eggs
(B) amastigotes
(C) promastigotes
(D) epimastigotes
(E) trypomastigotes

471. Infection of the men most likely involved contact with

(A) rodents
(B) mosquitoes
(C) sandflies
(D) ticks
(E) lice

472. Which of the following is the most appropriate confirmatory diagnostic test?

(A) blood smear
(B) CSF smear
(C) cultivation of synthetic agar
(D) hypersensitivity skin test
(E) growth inhibition test

473. An effective drug used in the treatment of the disease is

(A) amphotericin B
(B) nystatin
(C) flucytosine
(D) imidazole
(E) quinacrine

ANSWERS AND TUTORIAL ON ITEMS 469-473

The answers are: **469-C; 470-B; 471-C; 472-D; 473-A**. The disease experienced by the men is leishmaniasis. It is a zoonotic infection found in several tropical and subtropical regions of the world. The disease is transmitted through the bites of the sandfly. The men most likely were exposed to Leishmania braziliensis. L. tropica and L. mexicana has also been recovered from that part of South America. Typically the symptoms appear 3-12 months after the initial parasitic infection. A gradual progressive illness follows culminating in high fevers, weight loss, enlarged lymph nodes and enlarged spleen. The exact course of the infection varies slightly with different parasitic species.

The organisms are taken up by sandflies as they feed on natural hosts such as burrows and horses. Amastigotes are ingested which change into the flagellated promastigote form and multiply in the insect. During the feeding on humans, the promastigotes are injected. Then they multiply in the cytoplasm of macrophages as amastigotes. Amastigotes (Leishman-Donovan bodies) are easily observed in stained smears. L. braziliensis infection causes a strong cell-mediated immune response which has been used as a confirming diagnostic skin test. Other species of the organism such as L. tropica do not elicit a similar host cell response.

Amphotericin B is effective and widely used in treating the disease. A major concern is the likelihood of secondary bacterial infections resulting from the open lesions. Antibiotics are used to limit such infections.

Items 474-477

Twin 9 year-old sisters visited their aunt and uncle who lived on a small vegetable farm. During the visit the uncle showed the girls his pet pigeons which were kept in a large pen behind the house. They were allowed to play with some of the tame birds. A month following the visit one of the girls developed lingering headaches, dizziness and a mild fever. The symptoms progressed and as she experienced seizures. A CT scan showed inflamed meninges. Cerebrospinal fluid (CSF) showed a high protein level. A Gram stain of the CSF was negative for bacteria, but streaked on a blood agar plate produced large creamy-white mucoid colonies. The colony picked from the agar plate showed large circular structures with clear halos when mixed with India ink.

474. The organisms observed in the India ink wet mount preparation is consistent with a preliminary diagnosis of

(A) Candida albicans
(B) Blastomyces dermatitidis
(C) Cryptococcus neoformans
(D) Histoplasmosis capsulatum
(E) Coccidioides immitis

475. The girl may have been exposed to the organism

(A) during her normal daily schedule
(B) while playing with the pet pigeons
(C) through an infected playmate or adult
(D) by exposure from a pet dog or cat
(E) from playing in soil

476. The large capsule which surrounds the organism is composed of

(A) lipids
(B) nucleic acids
(C) proteins
(D) polysaccharides
(E) teichoic acid and cholesterol

477. A drug shown effective in the treatment of this infection is

(A) amphotericin B
(B) nystatin
(C) cycloheximide
(D) imidazole
(E) quinacrine

ANSWERS AND TUTORIAL ON ITEMS 474-477

The answers are: **474-C; 475-B; 476-D; 477-A**. The one twin is suffering from a meningitis caused by the fungus, Cryptococcus neoformans. Several types of birds, especially pigeons in the U.S., may carry the pathogen. The animals do not appear to experience any disease. Humans become infected through inhalation of dried bird droppings.

Microscopic examination of CSF from a patient with cryptococcal meningitis will frequently show large undistinguished structures which might be confused with lymphocytes due their large size of 4-7 μm. Gram stains are not very revealing due to the thick polysaccharide capsule which surrounds each organism. The India ink wet mount smear of CSF will show the cocci surrounded by a large halo caused by the capsule. Latex agglutination tests also are available which detect antibody directed against the organisms.

Candida albicans, Blastomyces dermatitidis, Histoplasmosis capsulatum, and Coccidioides immitis do not share the distinctive morphology and capsule characteristic of Cryptococcus neoformans.

Amphotericin B is effective in treating systemic disease caused by Cryptococcus. Another drug used against the fungi is flucytosine, however drug resistant colonies frequently emerge.

Item 478-483

Several compounds have been used successfully in the treatment of fungal infections. The mechanism of action of different compounds are not always the same. Match the following antifungal compounds with their modes of action and routes of administration.

(A) Nystatin
(B) Griseofulvin
(C) Amphotericin B
(D) Flucytosine
(E) Miconazole
(F) Potassium iodide
(G) Ketoconazole

478. A polyene which binds to ergosterol, resulting in membrane disruption; administered parenterally

479. A polyene which binds to ergosterol, resulting in membrane disruption; administered topically

480. Converted into 5-fluorouracil which disrupts protein synthesis, administered orally

481. An imidazole which interferes with ergosterol incorporation into fungal membrane lipids; administered topically.

482. Disrupts the microtubules of the mitotic spindle, interfering with cell division; administered orally

483. An imidazole which interferes with ergosterol incorporation into fungal membrane lipids; administered orally

ANSWERS AND TUTORIAL ON ITEMS 478-483

The answers are: **478-C; 479-A; 480-D; 481-E; 482-B; 483-G**. The different fungal agents respond differently to various antifungal compounds. This is due to a combination of factors such as the site of the infection (superficial, deep tissue or CNS). In addition, the variability of stages in the life cycle means that they might respond differently to antifungal agents. Similarly, different compounds have variable toxicity and solubility which alter their effectiveness. Many produce pronounced side effects which must be tolerated due to the limited alternative drug choices.

Amphotericin B (C) is a widely used antifungal compound which can be administered parenterally. It is a polyene which binds with the ergosterol found in the cell membrane of fungi. The binding results in weakening and rupture of the membrane. It has low aqueous solubility and must be administered intravenously as a colloidal suspension. It is effective against almost all fungi. It effectiveness has been increased by incorporating it into liposomes.

Nystatin (A) and miconazole (E) are applied topically. Nystatin is also a polyene and functions similarly to amphotericin B; however, its low solubility limits it to topical applications. Miconazole belongs to the imidazole group of compounds. It interferes with the synthesis of fungal cell membranes by blocking the incorporation of ergosterol into the cytoplasmic membrane.

Ketoconazole (G), griseofulvin (B), flucytosine (D) and potassium iodide (F) are all administered orally. Ketoconazole prevents the incorporation of ergosterol into fungal cytoplasmic membranes. It is less toxic than amphotericin B. Griseofulvin blocks the microtubule of the mitotic spindle, thus interfering with cell division. It is easily absorbed by the body and concentrated in different layers of the skin. Flucytosine in converted by cellular enzymes into 5-fluorouracil. The latter compound is incorporated into fungal RNA resulting in a disruption of protein synthesis. Fungal mutants resistant to flucytosine are common, limiting the effectiveness of the drug. Potassium iodide is only effective against Sporothrix schenckii. The mechanism of action is not well understood.

<u>Items 484-487</u>

The 35 year-old recipient of a kidney transplant developed severe chest congestion, coughing and pain. Antibiotic therapy did not appear to relieve the symptoms. An X-ray showed the presence of a large circular ball in the right lung. A biopsy showed tissue infiltrated with eosinophils and permeated with segmented branching hyphae. A specimen plated on nutrient agar produced a large fluffy gray colony that over grew the plate in 3 days and subsequently formed a puffy black-center colony.

484. The mold recovered from the patient belongs to the genus

(A) <u>Mucor</u>
(B) <u>Aspergillus</u>
(C) <u>Cryptococcus</u>
(D) <u>Rhizopus</u>
(E) <u>Histoplasma</u>

485. The dark appearance of the colony is due to pigment produced by

(A) mycelia
(B) conidia
(C) macroconidia
(D) microconidia
(E) sporangiospores

486. Infections of the organism, especially in immunocompromised patients, most often occur in the

(A) respiratory tract
(B) circulatory system
(C) urinary tract
(D) central nervous system
(E) skeletal system

487. The most effective drug for the treatment of this infection is

(A) cycloheximide
(B) imidazole
(C) amphotericin B
(D) nystatin
(E) quinacrine

ANSWERS AND TUTORIAL ON ITEMS 484-487

The answers are: **484-B; 485-B; 486-A; 487-C**. The mold isolated from the patient belongs to the genus Aspergillus. The morphology which includes branched septated hyphae, puffy mycelial growth and dark pigmented conidia distinguish the different Aspergillus species from the Mucor species. Several species of the organism are known to infect humans. These include Aspergillus fumigatus, the most virulent, followed by A. niger and A. flavus. They differ in morphology and antigenic profiles.

Most human infections of Aspergillus involve colonization of the respiratory tract. The mycelial growth progresses rapidly and may result in a concentrated or focal center of growth resulting in the development of a fungal ball. Biopsy material will yield the organism which is easily cultured on a variety of culture media. On agar plates, the mycelia quickly cover the entire plate with a puffy growth which develops a dark color as the fruiting bodies (conidia) mature.

Amphotericin B is the drug of choice in infections of Aspergillus. Advanced cases, such as those involving the presence of fungal balls, may require surgery in addition to chemotherapy.

Item 488-498

Knowledge of the morphology of molds and yeasts is important for accurate diagnosis and initiation of appropriate therapy. Match the labeled structures in Figure 4.3 with the most appropriate description in the items below.

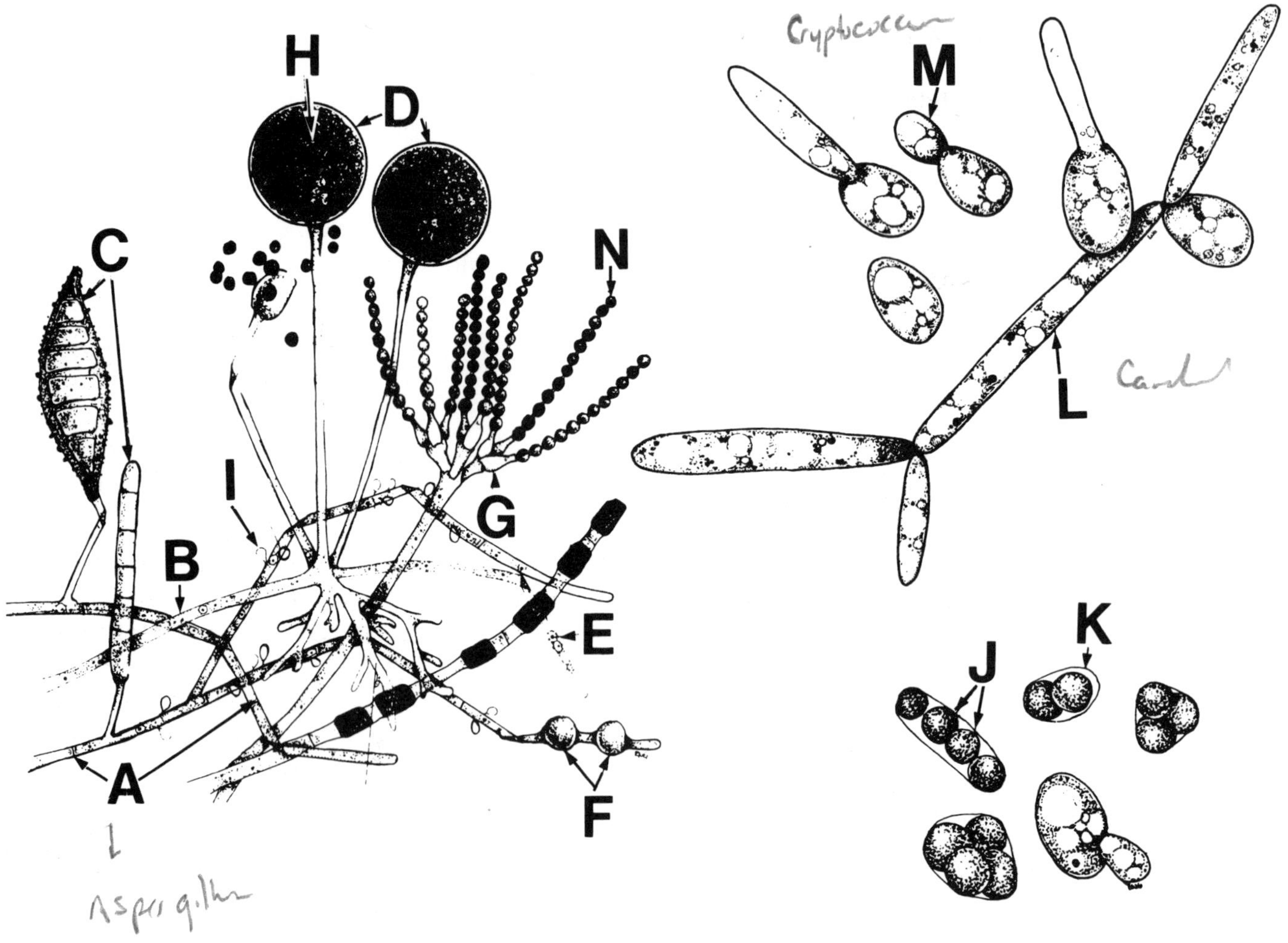

Figure 4.3

488. Bud or blastoconidia of Cryptococcus sp.

489. Hyphae of Aspergillus sp.

490. Pseudohyphae of Candida sp.

491. Sporangia of Mucor sp.

492. Hyphae of Mucor sp.

493. Macroconidia of Histoplasma sp.

494. Conidia of Aspergillus sp.

495. Ascospores of Candida sp.

496. Conidiophore of Aspergillus sp.

497. Ascus of Candida sp.

498. Sporangiospores of Mucor sp.

ANSWERS AND TUTORIAL ON ITEMS 488-498

The answers are: **488-M; 489-A; 490-L; 491-D; 492-B; 493-C; 494-N; 495-J; 496-G; 497-K; 498-H**. The structures illustrated in the composite figure of yeasts and molds represent forms observed from cultured isolates.

Ascospores (J) are the sexual spores of certain yeasts such as Candida. The ascus (K) is the thin case or covering which encloses 4 or 8 ascospores. A bud (M), also called the blastoconidia, is the asexual stage of yeasts and fungi. It is a smaller outgrowth from the parent cell.

Chlamydoconidia (F) are the sexual fungal stage. They develop within the hypha causing expansion of the hyphae walls. Conidia (N) are the asexual fungal stage which develop into fruiting structures not enclosed by a casing. The conidiophore (G) is the stalk-like structure on which conidia join the fungal hyphae. Macroconidia (C) are large aligned conidia which resemble segmented sausage-shaped structures in some fungi. Microconidia (I) are smaller bleb-like extensions along the supporting hyphal wall. Nonseptate hyphae (B) lack separations between nuclei within the hyphae resulting in one large multinucleated cell.

The nucleus (E) is membrane-enclosed and clearly distinguished within the hyphae. Pseudohyphae (L) are larger bud-like extensions interconnected by the smaller budding structures. Septate hyphae (A) have partitions between nuclei. The presence of these defines a fungal group. Sporangia (D) enclose the sporangiospores (or endospores) of the fruiting body. Sporangiospores (H) are the sexual stage of fungi encased in the sporangium.

Items 499-507

A young man returning to the U.S. from working several years in the Caribbean underwent a routine physical examination as part of an insurance policy application. The examination revealed an enlarged liver. The patient reported a history of abdominal pain and some diarrhea. Occasionally he noted blood in his stools which caused him to fear some form of cancer. A stool sample was collected and revealed large, 50x150 μm, oval-shaped eggs with a pointed spur.

499. The history and clinical picture of the patient are consistent with a preliminary diagnosis of

(A) giardiasis
(B) trichomoniasis
(C) trypanosomiasis
(D) leishmaniasis
(E) schistosomiasis

500. An important intermediate in the development of the organism are

(A) fish
(B) cattle
(C) snails
(D) rodents
(E) birds

501. The infection caused significant changes in the

(A) skin
(B) bowel mucosa
(C) central nervous system
(D) cardiovascular system
(E) skeletal system

502. Early stages of the infection are marked by growth of the organisms in the

(A) brain
(B) heart
(C) spleen
(D) liver
(E) lymph nodes

503. A strong inflammatory response is produced during the course of chronic infection due, in part, the presence of large concentrations of

(A) eggs of parasites
(B) dead adult parasites
(C) live adult parasites
(D) antigens excreted by adult parasites
(E) host-induced antigens

504. The adult stage survives the host cell immune response, in part, due to

(A) immunosuppression on the host
(B) the inaccessible location of the parasites
(C) masking of the parasite in host antigens
(D) encystment of the parasite
(E) antigenic variations of the surface of the parasite

505. An effective drug for the treatment of this infection is

(A) cycloheximide
(B) imidazole
(C) amphotericin B
(D) nystatin
(E) oxamniquine

506. All of the following measures are effective, or show the potential of being effective, in controlling the spread of this disease **EXCEPT**:

(A) improved sanitation
(B) control of snail population
(C) proper treatment and disposal of human sewage
(D) aggressive rodent control measures
(E) development of vaccines

507. The Katayama syndrome is characterized, in part, by

(A) conjunctivitis and arthritis
(B) encephalitis and elevated immunoglobulin levels
(C) endocarditis and elevated immunoglobulin levels
(D) conjunctivitis, urethritis and pneumonia
(E) elevated IgE and conjunctivitis

ANSWERS AND TUTORIAL ON ITEMS 499-507

The answers are: **499-E; 500-C; 501-B; 502-D; 503-A; 504-C; 505-E; 506-D; 507-B**. Infections caused by blood flukes such as the schistosomes remain prevalent in many parts of the world. The people of the Caribbean and South America have a high incidence of these infections. Schistosoma mansoni is the most widespread of the schistosomes in these regions (Figure 4.4). Maintenance of the organisms in nature is dependent upon passage through snails. Infected pools are often the point of infection. The organisms enter through the skin and migrate to the liver. During this early stage of the disease, the patient may experience fever, headaches and long-lasting abdominal pain. Each adult produces many eggs which stimulate the formation of immune complexes. These eggs play a major role in the pathogenic process.

Adult schistosomes evade the immune response, in part, due to the masking effects of human serum antigens. Many organs of the body are affected. The bowel mucosa often becomes inflamed and hemorrhagic, causing bloody stools. The Katayama syndrome is characterized by encephalitis, peripheral eosinophilia and elevated IgM, IgG and IgE.

Oxamniquine is effective in treating the disease. Other drugs such as praziquantel and metrifonate also show effectiveness against different species of Schistosoma. The spread of this disease can be limited through the development of effective vaccines, proper hygiene and sewerage disposal and snail control.

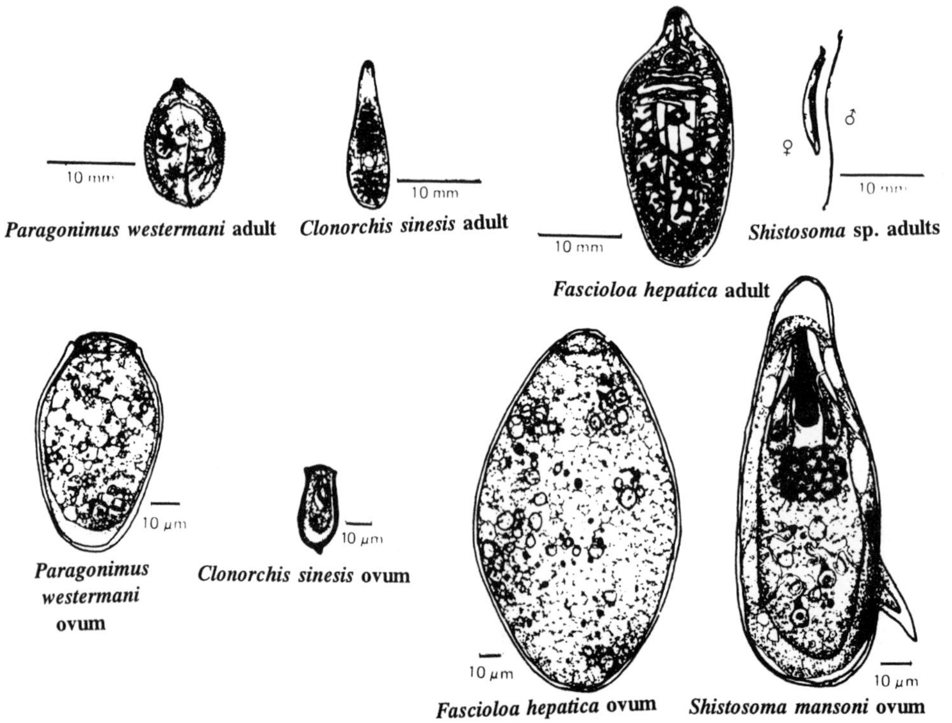

Figure 4.4

● Brief Review of Microbiology

- Bacteriology
- Virology
- Immunology
- Parasitology and Mycology

● Questions Formatted Like Current National Board Exams

- Clinical Scenarios
- Diagrams
- Problem-Based Questions
- Extended Matching Questions

J&S

J&S PUBLISHING COMPANY, INC.

ISBN 0-9632873-2-X